Estuaries of the World

Series Editor
Jean-Paul Ducrotoy, The University of Hull, Hull, United Kingdom

Estuaries are amongst the most endangered areas in the world. Pollution, eutrophication, urbanization, land reclamation; over fishing and exploitation continuously threaten their future. The major challenge that humans face today is managing their use, so that future generations can also enjoy the fantastic visual, cultural and edible products that they provide. Such an approach presupposes that all users of the environment share views and are able to communicate wisely on the basis of robust science. The need for robust science is pressing. Over the last decade there have been numerous advances in both understanding and approach to estuaries and more and more multidisciplinary studies are now available. The available scientific information has come from a multiplicity of case studies and projects local and national levels. Regional and global programs have been developed; some are being implemented and some are in evolution. However, despite the rapidly increasing knowledge about estuarine ecosystems, crucial questions on the causes of variability and the effects of global change are still poorly understood. Although the perception of politicians and managers of coasts is slowly shifting from a mainly short-term economic approach towards a long-term economic – ecological perspective, there is a need to make existing scientific information much more manageable by non-specialists, without compromising the quality of the information. The book series includes volumes of selected invited papers and is intended for researchers, practitioners, undergraduate and graduate students in all disciplines who are dealing with complex problems and looking for cutting-edge research as well as methodological tools to set up truly transversal science and technology projects, such as the restoration of damaged habitats.

More information about this series at http://www.springer.com/series/11705

Charly Machemehl • Olivier Sirost •
Jean-Paul Ducrotoy
Editors

Reclaiming and Rewilding River Cities for Outdoor Recreation

Springer

Editors
Charly Machemehl
University of Rouen
Mont Saint Aignan Cedex, France

Olivier Sirost
University of Rouen
Mont Saint Aignan Cedex, France

Jean-Paul Ducrotoy
Institute of Estuarine and Coastal Studies
The University of Hull
Hull, UK

ISSN 2214-1553 ISSN 2214-1561 (electronic)
Estuaries of the World
ISBN 978-3-030-48708-9 ISBN 978-3-030-48709-6 (eBook)
https://doi.org/10.1007/978-3-030-48709-6

This Springer imprint is published by the registered company Springer Nature Switzerland AG.
The registered company address is: Gewerbestrasse 11, 6330 Cham, Switzerland

Contents

Introduction: Reclaiming and Rewilding River Cities for Outdoor Recreation

Charly Machemehl, Olivier Sirost, and Jean-Paul Ducrotoy

Keywords

Urbanisation · Estuaries · Ecological restoration · Outdoor recreation

The concept of utilitarian nature has a long history, similar in scope to that of *Homo sapiens* (Lévêque and Mounolou 2008). The development of agriculture and the domestication of farmed species (deemed "useful") bear witness to this. In a global anthropocentric system of nature management that was instituted between the sixteenth and nineteenth centuries, the place of river towns quickly became central, as mentioned by Jean Brunhes in his human geography. Human societies are known to develop their own perception of nature. Nowadays, the village and its gardened extensions have become planetary, as we are reminded by Georges Guille-Escuret (1989) and Gilles Clément (2006).

In the context of anthropological exploration, the nineteenth century discovered that some societies collected things from the wild (through hunting, fishing and gathering) for pleasure, as Lucien Febvre mentions in his summary. At the same time, folklore surveys conducted in Western Europe (Van Gennep 1981) show that rural occupations based on the exploitation of flora and fauna are disappearing, leaving space, in the same range of activities and calendar rituals, for leisure pursuits. Picnics on the grass, pleasure trips, extended by camping play a large part in these dynamics. But already we note that the hygienist project of exiting city centres towards the outskirts is not exclusively urbaphobic (Baubérot and Bourillon 2009). A whole sociology on the rewilding of urban centres for the *Homo ludens* of the nascent twentieth century came into being. The grass covering the tarmacadam described by Walter Benjamin to designate the urban promenade, the erogenisation of the urban as a sensory experience as per Georg Simmel or yet again the turf playing field analysed by Jane Addams in Chicago tell us of the underground natural part of global metropolises. In the wake of such changes, the edges of the water became landscape and adventure playground. Frédéric Delaive showed that eighteenth century romantics gushed forth about amorous boating. In the second half of the nineteenth century boating became increasingly sportive with the development of nautical companies, canoes, rowing boats, then kayaks and motorboats that animate the urban landscape. The animation was not only sporty, but also focused on a bourgeois local hygienism with landscaping and gardening on towpaths, rambling and cycling on reserved paths, bathing establishments, riverside dance-cafés that inspired bucolic literature (Proust, Maupassant and Flaubert). These new endeavours fostered the development of the impressionist school to whom "luncheon on the grass" became the flagship theme of these rivers reclaimed for leisure pursuits. The elites enjoyed the private use of dedicated pontoons, boat garages, gardens and cottages, which became an integral part of the family "domus", as testified by the building of villas on the very edge of the river, that offered an urban holiday. For the popular classes, the magazine *Au bord de l'eau* (started by T. Burnand) and enthusiasm for line fishing showed a substantial appropriation of the river town. Beyond harvesting and picking wild plants and animals, a sociability of collecting emerged, together with *apéritif* drinks and decadent daytime napping that still imprints the imagery of outdoor leisure. Between the wars, two worlds rubbed shoulders in manufacturing the water city: the social elites who transposed the idyllic life of the romantic landscape onto the city

---

With support from GIP Seine-aval, FR SCALE CNRS.

C. Machemehl (✉) · O. Sirost
CETAPS EA 3832, Université de Rouen, Normandie Université, Mont Saint Aignan Cedex, France
e-mail: charly.machemehl@univ-rouen.fr; Olivier.sirost@univ-rouen.fr

J.-P. Ducrotoy
Institute of Estuarine and Coastal Studies, The University of Hull, Hull, UK
e-mail: Jean-Paul.Ducrotoy@hull.ac.uk

C. Machemehl et al. (eds.), *Reclaiming and Rewilding River Cities for Outdoor Recreation*, Estuaries of the World,
https://doi.org/10.1007/978-3-030-48709-6_1

(for example, the Bois de Vincennes on the edges of the Marne, as S. Beaudouin showed), and the visiting population crowding in a densified habitat, inland from the river. However, the multiplication of nautical festivals and, at the other end of the social scale, hawkers and pedlars encouraged a social mixity in leisure activities.

With the rise in political ecology (Moscovici 2002; Bozonnet and Jakubec 2000), plans to reclaim the towns expanded. Urban projects promoted social interaction in greened spaces. Prior to this, utopias such as garden cities, community gardens, eco-friendly farms, floating habitats and living homes (Willemin 2006, 2008) provided experience feedback that enabled the conception of urban regeneration.

In the experience of the Colonial Settlements in the United States, a community of interests emerged between social reformers and conservationists. The connections between Jane Addams' Hull House, John Muir's nature parks, Luther Halsey Gulick's YMCA recreational activities or the ethical culture of Felix Adler led to a general policy of urban reclamation combining wildness (national parks and wilderness areas) and leisure playgrounds, urban parks. In England as in Germany, outdoor movements and youth organisations proposed the same continuous to and fro between town and country.

From the 1880s, the hygienists and engineers driving social reform saw the city as a living organism that escaped them and that the ecologists described as a parasite. After much resistance, and despite the search for new uses, the urban excreta that had been contained for use by industry and spread on the fields for agriculture were abandoned to the rivers. Waterways running through towns became open sewers into which all types of waste, sludge and other mineral substances were dumped (Barles 2005). As early as the 1950s, the American metropolises proposed a programme of city greening while Germany continued to promote the ideology of the urban garden. It was also during this period that industrial wasteland was born, with the endless cessation of industrial production. The obsolescence of the machines, along with the delocalisation of industries, created a no man's land in which the landscapers intuited a potential playground (Chaline 1999). The general public's interest in this sensitivity to rewilding and reclaiming of the town centres for the people for leisure would be amplified after the hippie community experiences. These multiple threads of inheritance shed new light on the green engineering of which France would become of the cantors, with its young Ministry for Nature Conservation and its decentralisation policies. From then on the occasion for repairing the ecosystems, known as "restoration", advanced hand in hand with the urban planning projects, with the questioning of the dimension of ecological resilience and that of services rendered by the ecosystems (Clewell and Aronson 2010).

The scientific ecology underpinning such an approach was a young science that was only to gain its credentials in the second half of the twentieth century. For the scientific ecologist, the landscape comprised animal and plant communities distributed and interacting within a given geographic location: the ecosystem. Under this definition, the human dimension is de facto integrated. For not only is it impossible to consider a habitat without taking into account physical elements such as the relief of the land, the water cycle, etc., but it is also impossible to ignore the intervention of human societies that changes the frequency and/or intensity of certain factors such as, for example, artificial lighting, not to mention chemical pollution. The human activities that cause these unbalances result from sociopolitical choices based on various models depending on the cultures that are intertwined from the local to the global. Like other living creatures, humans affect their environment and try to adapt to it. Nowadays, the scale of perturbations and transformations imposed on ecosystems, and therefore on the landscapes we perceive, is increasing (Ducrotoy 2010). In the vein of the PIREN*[1] or SAUM** programmes at the end of the 1970s, French environmental science research examined the social demand for nature faced with the explosion of new towns and extensive urban spread (Lévêque and Mounolou 2008). The staggering progress of biology and the expansion of its domain of investigation characterise this switch to nature becoming worthy of the attention of the engineers. The first experiments to try to recreate natural ecosystems were carried out in the first half of the twentieth century by pioneers who were the first to envisage an ecosystem as an entity (Leopold 1949). There were multiple, complex reasons behind this revolution: a mixture of scientific and historical curiosity, an aesthetic interest but also a certain nostalgia. From the scientific point of view, respect for the old and established cannot be eluded with the notion that a "climax community" is ideally a preferential ecological combination of organisms endowed with the distinctive qualities of stability, harmony and a capacity for self-organisation. The concern for what was later to be called a "natural habitat" remained discreet until the end of the twentieth century. The idea of recreating a complete ecosystem or even an ecological community or a landscape, with all its attributes and processes, was long considered utopian (Jordan and Lubick 2011). For the public, these developments gave rise to the opening up of spaces hitherto condemned, often insalubrious or, at least, spoilt and degraded. The leisure society and its sports activities, endowed with an arsenal of equipment, took them over as early as the 1970s and the term then used was "restoration".

[1] *PIREN SEINE: *Programme Interdisciplinaire de Recherche sur l'eau et l'environnement du bassin de la Seine* **SAUM: Schéma d'Aptitude et d'Utilisation de la Mer.

By way of ecological restoration, engineering offered three possibilities: restoration *stricto sensu* (historical trajectory), rehabilitation (repair) and reassignment (new uses).

In its simplest expression, the restoration of ecosystems can be defined as the set of initiatives taken to improve an ecosystem, enabling it to achieve a better status, whatever this may be (Livingston 2006). This definition does not imply a return to a previous state, but only concerns the re-establishing of certain ecological functions. A definition now considered more standard mentions the procedures used to re-establish, depending on the degradations caused by human activities, a habitat or an ecosystem considered viable as much in its structure as in its functioning (Elliott et al. 2007). This is close to rehabilitation. In summary, restoring an ecosystem would consist in re-establishing the functions lost by this ecosystem, or replacing certain functions considered superfluous. In a radical way, we could decide to replace the affected estuary ecosystem (or certain habitats) by another ecosystem that would fulfil the functions referred to. This would involve reassignment of the territory concerned. The (re)-creation of habitats could, therefore, lead to opposite results. In the case of restoration, this could result in a system that has kept and even gained in estuarine characteristics; on the other hand, the river valley or estuary would no longer exist as such (Ducrotoy 2018).

When deindustrialisation began, it posed the question of how to convert urban and port wastelands, and of a partnership between industry and the wild (Rosenzweig 2003). As part of the collective utopia of well-being and quality of living spaces, the river city became an ideal model of a vast undertaking to reclaim through leisure, combining the technical questions of the rewilding of the environments. The metropolises had to have their nature, which were conceived of in terms of recreational services (Cronon 1991).

In the 1980s, the traditional activity of rivers, the transport of goods and passengers, started to decline. A consequence of economic upheavals, the port facilities (wharfs, access ramps, cranes and docks), the road and railway networks were transformed into wasteland (Chaline 1999; Collin 2001, 2007; Baudouin and Collin 1996).

These vast industrial zones that had been the bases of urban development became under-used and depreciated. Reoccupying them constituted a challenge for urban policies and mobilised public and private stakeholders within the scope of major projects (Bravard and Clémens 2008; Merle 2009; Saunier 2011). From then on the landscaping initiatives endorsed the social and economic transformations. They enshrined the shift to a service-based labour economy and also the essential place of leisure activities in our society (Dumazedier 1962; Corbin 1995). Thus, the landscape development projects were part of wider projects aimed at the reclaiming of river areas by the population. Functional space that served as a navigation channel was now appreciated for its landscape and aesthetic qualities that made it the preferential setting for fun, recreational and hedonistic activities. The town was refashioned by the desire for the shore (Corbin 1988). The restoration projects aimed at the degraded habitats took back possession of the wasteland areas. Their purpose was to create a bond between river and town, which the population had been deprived of, in a context of the development of the water and the massification of leisure activities. The river town was, in fact, a place where new ways of living together and being a society were invented.

Placing an environmental restoration project in a scientific perspective implies the application of the fundamental principles of ecology. However, due to the popularity of certain concepts such as biodiversity and biological productivity, the definition and use of the dedicated terms may be abused and mask the real issues involved in the development in question (Ducrotoy and Yanagi 2008). For example, for the general public and the politicians, biodiversity is the basis of robust and productive ecosystems, a fact that has never been proven scientifically. Recently, Elliott and Quintino (2007) pointed out the paradox of the quality of the estuaries, where a specific rudimentary richness supports a high level of production and assures the stability of a highly dynamic system.

Former harbour revitalisation projects were part of a global movement that concerned just as much river as coastal towns (Chaline 1999). Baltimore in the Patapsco River estuary and Boston in Massachusetts Bay were among the avant-garde. Back in the sixties, these cities repurposed their harbour areas. The principle of mixed uses was given preference over zoning and it favoured the introduction of leisure facilities and activities. The waterfront took on a heritage dimension, in particular through the conservation of the harbour past. A second stage was completed in the 1990s. To the urbanistic rationale oriented towards "the back of the waterfront", therefore the revitalisation of the town centre, succeeded a rationale of "the front of the waterfront", giving preference to the aquatic element (Vermeersch 1998). In this second phase, greater importance was accorded to leisure activities. In degraded ecosystems, it was at this time that the principle of use and public access to the space became obligatory and the landscaping of parks included land planning. In Europe, riverbank revitalisation and repurposing projects in Liverpool and London began in the 1980s, followed by Amsterdam, Antwerp, Barcelona, Bilbao, Genoa and Rotterdam. In France, the process concerned Bordeaux, Lyons, Orléans, Paris, Rouen, Valence and Vienne. The decision to make the riverbanks accessible to the locals was shared by all the cities. It was expressed through the development of public spaces in the form of landscaped promenades, cycling tracks, playgrounds, gardens and parks in proximity to the water and bounded by the renovated, showcased quays (Lechner 2006).

In the river towns, leisure sports facilities were the principal means of re-appropriating the banks and habitats taken from the aquatic domain. Public policies and the reflections they stimulated sought to renew the bond between river, estuary and the inhabitants. In addition to hotel and catering facilities, shopping malls and boat coaches, the largest urban centres endowed themselves with sports enclosures destined for shows and sports, as well as marinas. Public spaces such as promenades, gardens, parks and open sports facilities also occupied an important place. Even in the most modest towns, the landscaping of the riverbanks included the creation of pedestrian or cycle paths, riverside stopovers, playgrounds, relaxation and picnic areas or signs pointing out remarkable architectural features or panoramas. Heritage buildings were destined to maintain and promote water-related traditions (fishing, shipyards, etc.) or to implement systems to protect nature. Initiatives intended to bring the population closer to the river became widespread. They included the rehabilitation of the landscape function of the urbanised river. The aim was to reinvent a way of looking at water and a physical, experiential and sensory relation to water.

The reclaiming of the shores and aquatic habitats transformed or degraded by sports leisure activities, encouraged by the Millennium Ecosystem Assessment (2003), was a response to four principal challenges. The first concerned the spatial dynamics of the practices and forms of recreational appropriations. The introduction of sports and leisure facilities implies a reflexion on the distribution of space and how they overlap to satisfy functional needs. While creating areas of freedom, landscaping induces certain types of practices and encourages certain forms of sociability. Leisure time gradually became a part of urban rhythms. In the light of the reflection of Hans Jonas (1979), we may wonder about this landscaping policy based on mass consumption. Does it not, in the end, aim to satisfy needs that are temporary and not essential?

Secondly, the constitution of new living spaces implies the building of a sustainable town. It renders necessary the depollution of the sites, landscaping compliant with the legislation and environmental standards. The link with nature is made more evident thanks to the presence of the water and the rewilding of the former production and storage sites and Non-places destined for transportation. The showcasing of the natural heritage is also prone to raising the awareness of the general public to environmental issues (via nature reserves, museums, zoos and parks). Quality of life and well-being are also presented as the stated objectives of the river towns. An extensive offer of leisure activities—entertainment, fun, sports and culture—favours personal development under many facets—physical, psychological and social fabric.

Thirdly, for the development of outdoor activities, wetlands and aquatic river and estuary milieus must recover their basic ecological functions. The intertidal zones are the most vulnerable due to the land needs of the ports, industry and agriculture. The neighbouring town where the players of this occupation reside are most often outside the valley or the estuary (where the harbours and industries are located) but most frequently occupy fragile peripheral areas. The priority for restoring them must be recognised and must be subject to a veritable consensus within the scope of an ecosystemic and interdisciplinary approach. Recent studies (Moussard et al. 2008) showed that in the estuaries of the Manche the basic functionality was still present, but was more or less failing. The priority was to reconnect fragile functional links between the alluvial plain and the axis of the river in its minor bed, leading to an improvement in the hydrological and biological processes. Such developments naturally had to be conceived of in a holistic spirit, incorporating restoration projects into an overall development scheme for the river and its estuary, projected over the long term. A long-term environment friendly vision based on a good knowledge of the past histories of the sites concerned, can enable scenarios to be built to promote local activities in harmony with the forced changes to ecosystems in the face of contemporary climate disruptions.

Fourthly, the image of the town is worked on through the urban planning projects which, in a global economy, seek to capture the stream of tourists and especially capital by endowing themselves with an appealing image. The tourism dimension is superimposed on projects primarily destined to adapting the functioning of the ecosystems to the marked changes in climate. The existing infrastructure and rigid administrative frameworks prevent the creation or restoration of vast expanses of wild spaces in the more artificialised areas of the wetlands. However, suitable, diligent planning, combined with creative design could be successful. By forging a positive identity, in the era of communication, the cities were trying to play a positive role in the flows, but this urban marketing used certain cultural, corporal and sports practices as a foil and pushed others aside, creating the risk of a homogenisation of cultures and lifestyles (Evrard et al. 2012).

These four challenges are addressed in the chapters of this book. Their authors have discussed the issues according to which outdoor leisure pursuits define places and reconfigure urban territories, manufacturing the river town. Thus in "the cities and their waterways" and by way of an introduction, Christian Lévêque proposes a genealogy of the urbanisation of rivers via their development and uses. He describes several movements. First of all, he explains how the locals make use of the water through fisheries, mills, agriculture and the exploitation of marshes are transmitted through folklore. A recreational and patrimonial tradition which today takes the form of a new agenda of celebrations of nature in the cities. Then he shows how, in the course of the 1970s, the protection of the cities from turbulent waters (floods, erosion, the transport of goods and passengers) turned towards the engineering of ecological restoration, showcasing the rehabilitation of the open sewers and the natural spaces that had been considered lost for the population (islands, mudflats, banks, etc.). Finally, he shows how, in a movement of urban regeneration, the city centres gravitate towards the edges of the water to be reclaimed, in particular, due to deindustrialisation that left vast wasteland reoccupied by the citizens as playgrounds. In these three instances, Christian Lévêque emphasises the impossible modus vivendi between ecological restoration and urban regeneration which are conceived of according to different spatial and time frames.

In his article, Jean-Paul Ducrotoy deals with the restoration of rivers and estuaries and nature-based activities posed as a departure point for re-thinking the necessary reconnection of humans to their living environments. He offers a very complete panorama of the environmental issues, from a comparative study of several estuaries; sediment management to facilitate navigation (the Elbe), flood prevention and port access (the Escaut), hunting and the development of tourism (the Somme) or compensating for the consequences of the extension of industry and harbours (the Seine). In the same way as the approaches through landscaping, sociology and anthropology, he defines new perspectives on restoration and on envisaging the consequences for the environment, people and society. Ultimately, a committed text that warns of a way of thinking and of devising public policies that are incapable of dynamically defining restoration projects centred on the human. He is essentially defending better sharing of power, which he believes can be achieved through the decentralisation and consistency of European action, and makes political commitment a necessary condition for overcoming the crisis of the estuaries. The crisis could be favourably resolved on the condition that a sustainable bond is forged between people and their waterways.

"Behind good ecological status, the quest to reconquer water territories", the article by Olivier Sirost and Charly Machemehl, is based on the comparison of the three French Atlantic estuaries; the Seine, the Loire and the Gironde. The authors underline the implications of good ecological status, as defined by a European framework law, in these estuaries which are highly distinctive in terms of spatial characteristics and social expectations. The latter turns out to be decisive for an understanding of the paths to restoration that cannot be separated from local social rationales.

Sylvie Miaux and Maxime Demers-Renaud wonder about the manner in which the developers think and conceive river spaces with the aim of enhancing and developing areas for recreational tourism. They show that urban promenades—physical, developed spaces—are redefined and mobilised in the light of health and sustainable development values for building social bonds and instigating an active lifestyle. Based on case studies from Québec, Bordeaux and Saragossa, the reflection examines the modes of production of the town, not only as a functional space but also as a cultural one. Nature plays a special role here. Projects bring into play the accessibility, appeal and staging of nature that is envisaged, depending on the case, as a prop, to promote a site or a town or, wild, a sensorial and experiential immersion.

In his contribution entitled "Outdoor leisure activities at odds with the city? Arcachon Bay and the Massif des Calanques", Ludovic Ginelli compares two meccas of nature located on the edge of the city. On the basis of a survey of users, clubs and managers of protected areas, he shows the discrepancies between the representations of the great outdoors and the experience of an environment permeable to the influences of the town and its social issues. Sports leisure activities deployed in the densified outskirts such as bow hunting, underwater hunting and sea kayaking are caught up in a bundle of contradictions. They embody the liberty of the fusional experience of nature in contrast to urban enclosure. But more recently these sports have also become the locus and subject of an eco-citizen debate and a naturalist participation involving the co-construction of a natural protected area. While these practices may come up against a sanctification of wild nature, they are more deeply a part of a porosity that tries to go beyond the urban–rural division and sketch new ways of managing protected spaces.

In a comparison of the Saint-Laurent estuary (Canada) and the Gironde estuary (France), Sarah-Jane Krieger invites us to examine the proposition: "On the conquest of wild nature . . . but what is meant by 'nature'?" From two quality surveys through interviews and observations requiring six months' immersion, Sarah-Jane Krieger questions how devotees of outdoor sports—sailing, riding, kayaking, rambling and cycling—represent and experience areas of wild nature that are subject to strong protection measures. Social-natural objects such as islands, paths, wildlife observation points, spots for gathering fruits and nuts, and wetlands reveal a desire for the absence of human activity in which to recharge your batteries, and, at the same time, nurture an urban-phobia in speeches and political stances. But paradoxically these immersive leisure activities are accompanied by increasingly heavy developments that are accepted as long as their artefacts remain invisible or at the very least in harmony with a mythical, original wildness. Without a doubt, we may read in these ways of introducing "wild beauties" into the urban environment, the object of symbolic and social investment for the city dweller.

Laura Verdelli and Noémie Humbert take us into the analysis of the interface between the city and the water via their text: "La Darsena di Milano: 'restoration' of an artificial urban aquatic milieu between citizen hopes and municipal plans". This mutation, affecting saturated urban areas, has been accompanying European cities for around thirty years. We can see in this a movement that draws its discourse and actions from the return of nature in towns (the concept of urban green belt and blue belt is particularly eloquent in France), the safeguard of cultural but also "natural" heritage, of the practice of sports, the development of outdoor leisure, and the promotion of a new "sustainable" image of the town. Behind this culture of eco-responsibility and eco-citizenship heterogeneous transformations of space are being woven. There is indeed a common thread running through these points of view, committing the local players and the population, labelling the projects... But in the end, the case study of la Darsena in Milan primarily shows the inadequacy of a green veneer. The deeper ecological debate is replaced by an eco-friendly landscaping supported by the expansion of urban leisure activities.

Mickael Attali and Natalia Bazoge take a look at the case of the Isère and the development of recreational activities around this fast-flowing river, which starts in the Alps (la Vanoise) and flows into the Rhône to the North of the city of Valence. They show that waterways were introduced into urban areas through leisure activities from the end of the nineteenth century. From a hostile territory, through outdoor sports, the river became a playground for exploring and relaxing, in the heart of the town. In particular between the wars, the waterways reposition themselves with respect to an older, more mobilising concern for the development of the mountains.

Damien Féménias, Olivier Sirost and Barbara Evrard take a look at the role and the signification of leisure activities in harbour areas, taking the reinvention of the quays of Rouen for example. The authors examine the articulation between public policies and their effects on the leisure pursuits on offer. Beyond this, they show that through leisure activities and access to nature a harbour neo-urbanity is built up, which bears witness inter alia to the resilience of a land heavily affected by deindustrialisation.

Sébastien Bourdin and Yann Rivoallan adopt a geo-prospective approach to question the future development of the Caen Peninsula (urban community of Caen at the sea) and recreational/sports activities. Drawing on the Guide Map adopted by the metropolis, they show the specific features of the operations to reconvert and revitalise their harbours and reflect on the potential for economic development within the scope of a river-land-sea continuum. The authors show the conditions under which the different facets of the project—urbanistic, social, environmental and mobilities—can constitute an opportunity for the attractiveness of the region, without forgetting to point out the risks.

Jean-Pierre Augustin invites us to a comparative reading of two different types of maritime life superimposed on each other, which partake of the social-spatial construction of the city of Bordeaux. In a text entitled: "Bordeaux's playful and sporty maritime life: a revolution of venues and activities", he analyses one movement based on work and trade, and the other related to leisure and sports. Since the 1960s, this double usage and development of space have been accentuating in favour of the recreational. By analysing the landscaping of the quays, spaces dedicated to sports practices and the structures built for leisure activities, the geographer shows how another meaning of space has grown through time and raises awareness once again of the developments of the past. Bordeaux has drawn on its economic and industrial heritage to free access to the water and its edges. Here we can read the stylistic codes of the twenty-first century city playing with the aesthetics—a new shared maritime phenomenon—of the water.

References

Barles S (2005) L'invention des déchets urbains. France: 1790–1970. Champ Vallon, Seyssel

Baubérot A, Bourillon F (2009) Urbaphobie ou la détestation de la ville aux XIXe et XXe siècles : actes du colloque réuni à Paris-12 Val-de-Marne, les 8 et 9 mars 2007 (vol. 1–1). Pompignac: Éd. Bière

Baudouin T, Collin M (1996) The revival of France's port cities in European ports. J Econ Soc Geogr 87(4). https://doi.org/10.1111/j.1467-9663.1998.tb01564.x

Bozonnet J-P, Jakubec J (2000) L'écologisme à l'aube du XXIe siècle. De la rupture à la banalisation? Georg, Genève

Bravard J-P, Clémens A (eds) (2008) Le Rhône en 100 questions. Zone atelier bassin du Rhône (ZABR), Villeurbanne

Chaline C (1999) La régénération urbaine. Presses universitaires de France, Paris

Clément G (2006) Où en est l'herbe? Réflexions sur le jardin planétaire. Actes Sud, Arles

Clewell A, Aronson J (2010) La restauration écologique. Actes Sud, Arles

Collin M (2001) Nouvelles urbanités des friches. Multitudes 6 (3):148–155

Collin M (2007) Nouvelles urbanités des friches de l'époque industrielle. In: Moulier Boutang Y (ed) Politiques des Multitudes. Éditions Amsterdam, Paris, pp 583–587

Corbin A (1988) Le territoire du vide: l'Occident et le désir du rivage (1750–1840). Aubier, Paris

Corbin A (1995) L'avènement des loisirs: 1850–1960. Aubier, Paris

Cronon W (1991) Nature's metropolis: Chicago and the Great West. W. W. Norton, New York

Ducrotoy JP (2018) Paysages naturels et activités humaines en Baie de Somme: Vers une anthropologie écologique

Ducrotoy JP, Yanagi T (2008) Tools and concepts on ecological quality of coastal and estuarine environments. Mar Pollut Bull 57(1–5):1–2

Ducrotoy J-P (2010) La restauration écologique des estuaires. Editions Tec et Doc, Lavoisier, Paris, 229 pp

Dumazedier J (1962) Vers une civilisation du loisir ? Ed. du Seuil, Paris

Elliott M, Quintino V (2007) The estuarine quality paradox, environmental homeostasis and the difficulty of detecting anthropogenic stress in naturally stressed areas. Mar Pollut Bull 54:640–645

Elliott M, Burdon D, Hemingway KL (2007) Estuarine, coastal and marine habitat and ecosystem restoration: confusing management and science. Estuar Coast Shelf Sci 74(3):349–366

Evrard B, Féménias D, Machemehl C, Letourneur O, Bouillon D, Sirost O (2012) Sequana: constructions sociales de l'estuaire de la Seine, rapport Seine-Aval 4, Oct 2012, 180 p

Guille-Escuret G (1989) Les sociétés et leurs natures. Armand Colin, Paris

Jordan R, Lubick GM (2011) Making nature whole: A history of ecological restoration. Society for Ecological Restoration, Island Press, Washington, DC

Lechner G (2006) Le fleuve dans la ville. la valorisation des berges en milieu urbain. (Centre de documentation de l'urbanisme, Éd.). Direction générale de l'urbanisme, de l'habitat et de la construction, La Défense, France

Leopold A (1949) A sand county Almanac. Oxford University Press, New York

Lévêque C, Mounolou J-C (2008) Biodiversité. Dynamique biologique et conservation. Dunod, Paris

Livingston RJ (2006) Restoration of aquatic ecosystems. CRC Press, Boca Raton, FL, 423 pp

Merle B (2009) L'estuaire de la Seine: un passé en commun, un avenir en construction. Éd. des Falaises, Rouen

Millennium Ecosystem Assessment (2003) Ecosystems and human well-being. Island Press, London

Moscovici S (2002) De la nature. Pour penser l'écologie. Métailié, Paris

Moussard S, Ducrotoy JP, Dauvin JC (2008) Gestion Globale de l'Estuaire de la Seine : de sa dégradation à la restauration de ses fonctionnalités écologiques. In: Houdet J (ed) Intégrer la biodiversité dans les stratégies des entreprises. Orée, Entreprises, Paris, pp 324–339

Rosenzweig ML (2003) Win-win ecology. Oxford University Press, New York

Saunier F (2011) Histoire d'une pionnière: l'agence d'urbanisme de la région du Havre et de l'Estuaire de la Seine, 1965–2010. Éd. des Falaises, Rouen

Van Gennep A (1981) Manuel de folklore français contemporain. Picard, Paris

Vermeersch L (1998) La ville américaine et ses paysages portuaires: entre fonction et symbole. L'Harmattan, Paris

Willemin V (2006) Maisons vivantes. Ed. Alternative, Paris

Willemin V (2008) Maisons sur l'eau. Ed. Alternative, Paris

Cities and Their Waterways

Christian Lévêque

Summary

Waterways have played a significant role in our economic development. The navigation of waterways for the transportation of passengers and freight was very active from antiquity till the beginning of the twentieth century. Much landscaping was done to facilitate the activity: levelling of thresholds, canals, locks and millstreams, etc. Cities and their ports grew up around strategic points. Waterways also provided craftsmen, then nascent industry with the engine power necessary for their activities, which was at the same time the source of much pollution. Gradually these uses regressed to make way for other, recreational uses in the nineteenth century such as boating, bathing and riverside dance cafés. The rise in awareness about the environment after World War II led to waterway ecosystems being given more consideration, in particular with, the fight against pollution and so-called restoration operations. In certain towns, the ports and industrial sites were redeveloped into spaces for recreation and entertainment, with operations which, under the guise of restoration, in reality created new business areas. City dwellers who sometimes dream of "wild rivers" are actually attached to these places steeped in history, which have witnessed the activities of the past, the memories of which are revived in folk events. A new urban identity is arising, anchored in the valorisation of a heritage with considerable identity value. A process that sometimes opposes the naturalist ecological restoration projects.

Keywords

Uses of waterways · Recreational areas · Landscaping · River heritage · Reconquer the banks

C. Lévêque (✉)
Directeur de recherche émérite, Institut de Recherche pour le Développement, Ville d'Avray, France

Introduction

> Water is to the urban what land is to the rural. (Guillerme 1990a)

> Relations between towns and rivers have always been agitated. While waterways have frequently, throughout history, conditioned the prosperity and growth of towns, river spates have always been a major constraint to which they have adapted more or less successfully. (Bravard 2004)

The development of numerous urban centres is closely associated with the waterways (Lévêque 2016, 2019; Beauchêne 2006). In addition to the supply of water, rivers enabled the transport of everything the town needed: wood for heating, coal, foodstuffs, building materials, and industrial materials. "*Without the Seine, Oise, Marne and Yonne, Paris would not be able to eat, drink or even heat itself comfortably*" wrote Fernand Braudel. Not counting the driving force of water, which was long the main source of energy for craftsmen and nascent industry. We could also add that the river was a convenient way to evacuate all the waste produced by the city... When waterways traverse the city, they are therefore once a resource, a tool and a threat during spates. By the activities it generates, and the dangers it creates, the waterway structures the town. Along the waterfront, warehouses and other harbour activities grow up. On the edges live those who work at the port, whereas the homes of the wealthy are situated farther away, often higher up. The result is cities split between the "top of the town" sheltered from rising waters and the nuisances of the port, and a "lower town" with its "lower districts" sometimes prone to flooding, home to a poor population, and a place of productive activities.

Before the French Revolution, the river was a place of intense activity. In addition to commercial traffic, fish was caught there and kept and commercialised in the ponds; clothes were washed in large laundry boats; horses were taken to drink and to bathe. And so lots of minor trades lived directly off the river: smugglers and small boats, water carriers, washers and carders of mattresses, boat

C. Machemehl et al. (eds.), *Reclaiming and Rewilding River Cities for Outdoor Recreation*, Estuaries of the World,
https://doi.org/10.1007/978-3-030-48709-6_2

scrappers, labourers carrying coal, "hookers" carrying wood for burning. These men and women in their hundreds were the people of the river, the water people.

Navigation and Ports

One of the main uses of waterways has been put to is navigation. When overland routes were rare and unsafe, rivers were preferential channels for the circulation of people and goods. Commercial exchanges have existed since Antiquity, whether for raw materials or local produce, such as, for example, the Italian wine that was imported via the Rhône for the consumption of the Gauls, or the wine produced in the Loire Valley and exported to northern Europe. These "roads which move" as described by Pascal were traversed by all kinds of craft. The development of river navigation was a real strategic challenge for the national economy from the age of the Gauls up until the advent of the railways in the nineteenth century. It is hard to imagine the bustling life of the ports in the old days, especially in the cities that had to be supplied continuously... At that time, inland navigation and river trading occupied a large population, and required the upkeep of a considerable cavalry for the towing of the boats (Le Sueur 2012).

The Kings of France had understood the strategic importance of river navigation, which is why they gradually brought this sector of activity under their authority. A Capitulary of Charlemagne placed the navigable rivers and tributaries under the authority of the Emperor. Subsequently, the power of the monarchy became stronger and was extended, with all activities undertaken on the waterways requiring authorisation from the King. As a result, the installation of mills and fisheries was subject to authorisation to avoid obstructing the flow. The purpose of the royal custodianship was therefore to ensure free circulation of people and goods within the kingdom, which was indispensable to economic development. Here, already, we find the idea that prevailed with the railways: ensure the circulation of production to connect the provinces and encourage trade.

The port is a developed space destined for the loading and unloading of goods and the embarking and disembarking of passengers. Situated in the heart of the town, the port was the nerve centre of urban life. It is the centre of many water trades, inland navigation and all the port activities. Of course, the towpaths ran through the cities with their processions of horses. Water coaches transported passengers from town to town. Between the sixteenth and the beginning of the nineteenth centuries, all the representations of the river banks in major towns, on prints and paintings, testify to this intense activity.

Until the eighteenth century, there were no storage areas in the Parisian ports, as the merchandise was sold on gradually as it arrived. The banks were cluttered with tethered boats for trading (charcoal, wood, wine and fruit), and there were lots of boats waiting in the nearby riverside villages, tethered along the banks, before docking in one of the Parisian ports. Throughout the nineteenth and twentieth centuries, due to the increase in traffic and the development of industrial activities, the ports moved further away from the city centres to reduce the disturbance they caused.

The history of a port is the reflection of a stratification of uses that have evolved in accordance with commercial and industrial activities, the progress in navigation and in the means of communication. When these activities disappear, the port collapses... This is what happened at the end of the nineteenth century, when the railways, then the road system, replaced river transport.

The Development of the Waterways for Navigation

Navigation faced major difficulties (Barrier 1990, Berthier 2009):

- The draught of the boats. Over time, the size of the vessels increased, requiring greater flow depths, which led to the deepening of the riverbed, for example a phenomenon still with us today. But in the towns, the river was contained in order to narrow the bed and increase the flow speed, which contributed to further deepening the bed. Then in the nineteenth century, stones were used on the banks to prevent erosion and to facilitate the landing of boats.
- "Water idleness", that is to say the period during which navigation was not possible, was a severe obstacle to trade. The economic apparatus of the city was closely dependent on the rhythm of the flow of the waterways, which turned the mills and decided whether boats could pass. The difficulties facing navigation at certain times of the year posed considerable problems for the supply of the towns. Large scale development work was undertaken to remedy this, including the correction of routes and deepening of the channel, containment of the bed, the creation of dams and locks, etc. These developments, which changed our waterways profoundly, are still present, despite a real antipathy for river transport. The river tourism that subsequently developed remained modest compared to the investments accorded it. Nonetheless, this developed environment is part of our heritage.
- Reloading. While boats were built to sail as far as possible up the waterways, the fact still remained that at certain times goods had to be unloaded and transhipped by land to the next waterway. As early as the seventeenth century, numerous canals were built to connect the river basins. The Briare Canal (1642), for example connected the Rhône basin to that of the Seine. The Loire and the

Rhône were connected by the building of the Canal du Centre in 1792. Currently, there is a whole network of canals connecting the various river basins. They are much used for river tourism (Miquel 1994).

In the second half of the nineteenth century, inland navigation regressed rapidly in the face of the railways. All that remained was, for a short time, the transport of passengers and some goods in town centres. The unused banks became abandoned wasteland and ill-frequented...

Cities and Flooding

Another factor that contributed to distancing the town from the waterways was flood prevention. In the nineteenth century, there were extreme flooding events that lingered in the memory.

The flood protection system we know today was largely inherited from the policies implemented in the past, the history of which allows us to trace the thread of the changes that occurred in the river landscapes. The multiple developments, when we think about them, were a kind of DIY, with frequent adjustments, like the dykes that were continually being built up further in the course of the nineteenth century in step with recurring contingencies. Still, these developments, often carried out jointly with those necessary for navigation, led to the increasing artificialisation of the river system: waterway straightened, banks concreted, bed deepened, etc.

Following the disastrous spates of the nineteenth century on the Garonne, the Loire and the Rhône, the defence against floodwaters was the reason for some of the more structural urban developments, such as quays to replace dykes, and serve first of all as ramparts against floods (Porhel 2012). The quays in stone improved the conditions of transportation and unloading, but surrounded and channelled the river, which lost its wild natural aspect. The quays replaced the shingle shores and the banks that up until then had been used for a number of activities. In many towns, channelling the river went along with the building of a system of upper quays on which the circulation routes were established, and, down below, of lower quays which long served for goods handling and haulage. The rise in land transport at the beginning of the twentieth century gradually led to the banks being occupied by the railway and road infrastructures. Symbolically, the quays isolated the river from the town. Access to the water was no longer direct; you now had to go down to the lower quay to reach the water. The disappearance of the houses along the edge of the river further accentuated this distancing of the waterways.

> Throughout the 19th century, State intervention played an essential role which removed all sociability from the water by confining it between walls and restricting access to it. (...) The river is a place of transit, transportation, trade, meetings, and also of prostitution. It is a lively place which is beyond the control of society and the economy. The State, on the pretext of combating flooding, removed the shingle shore in Paris, and replaced it with upper and lower quays, veritable military installations that cut the river off from the town, and enabled the taxing of commerce. (Guillerme quoted by Lechner 2006)

And there were several attempts to capitalise on the feedback of experience from the past to anchor current protection policies. But paradoxically, land ownership pressure tended to obscure the negative consequences of the flooding in numerous places. Perhaps people hoped that these events would not recur in the near future? Perhaps there was a lapse of memory in citizens who had not been confronted with the problem in the absence of a recent disaster? The fact remains that there is still a strong temptation to build in flood-prone areas.

The Era of Productivist Industrialisation: The Landscaped River

In the nineteenth century, technical resources improved considerably and several phenomena contributed to profoundly altering the relationship between town and waterway. First of all, the design of the suspension bridge made it possible to multiply the points of passage, enabling easier liaison between the two banks of a town. Between 1827 and 1853, eight of these structures were built on the Saône at Lyon. Fifteen bridges were built in Paris in 1870, a greater number than during all the previous centuries. And the design of metal bridges enabled the construction of viaducts for rail transport. So many realizations that would contribute to the growth in rail and road transport to the detriment of river navigation, which began an irremediable decline despite the arrival of steam power.

But due to the economic conditions, industry took over the banks, as much for reasons of access to transport as for supplying their boilers and rejecting their waste, which aggravated river pollution and usage conflicts. In medieval times the river already attracted skinners, tanners, papermakers, launderers, butchers, dyers... whose activities used a lot of water and at the same time were sources of pollution. For convenience, the waste was discharged directly into the current, causing conflicts with those who needed safe water for their activities, and the residents who were looking for clean water for domestic use. In the nineteenth century, the rampant industrialisation stimulated the development of vast industrial and logistics strangleholds in the major cities which contributed to further isolating the urban fabric from the river, which had gradually become a place that was not very welcoming for the townspeople. And, to feed the industrial ogre which devoured coal and raw materials, the locks were widened and the channels deepened to allow the navigation of increasingly large boats. And as room was needed for these new activities, ports grew up outside of the towns.

Furthermore, and for a variety of reasons (the development of hydroelectric power, the supply of the waterways when levels were low, or for fighting floods, irrigation of agricultural land, etc.) large numbers of dams were built from the end of the nineteenth century. Fairly quickly, the "small inland seas" they created were put to use for boating and bathing.

New Centres of Interest, New Uses. . .

There is nothing simple about the relations between societies and their waterways. In the nineteenth century, while the town was becoming removed from the waterway which was of less importance for transport and polluted by industry, citizens began to find in it a new appeal and new uses, this time recreational. The term "refound river" was sometimes used. However, it would be fairer to say that the rise in leisure activities created new needs that the river was one way of meeting. This was because, during the nineteenth century, the free time that had been the sole preserve of the wealthy classes gradually became a privilege shared, albeit parsimoniously, with the popular classes. A society of leisure was emerging, partially favoured by the new modes of transport such as railway and road. A good number of townspeople began to spend their free time on a Sunday in proximity to a waterway. In Paris, with the first local railway between Paris-Le Pecq inaugurated in 1837, the still relatively deserted banks of Seine which became accessible around the capital, and soon Normandy followed. The end of the nineteenth century was the golden age of the riverside café, boating and the impressionists. It was also the time of weekends and leisure activities with the growth of recreational uses of the waterways: bathing, boating, fishing, etc. But most often, thanks to the railways, people travelled outside of the towns to find banks that were still "natural" devoted to leisure activities. Having known a prosperous period, these recreational activities partially abated, or were transformed, at the same time as the industrial activities strongly criticised for the pollution they generated continued to decline.

The use of the waterways for leisure activities is a process anchored in the second half of the nineteenth century, when urban dwellers did not have access to the sea shore and found a substitute: the river (Guillerme 1990b; Berthier 2009).

Bathing

In the seventeenth century, the Parisians bathed in the Seine in large numbers, in spite of the filth it could carry around. In the eighteenth century, the first bathing establishments appeared along the banks of the Seine in and around Paris. Places for bathing quickly improved in the nineteenth century to offer more comfort (individual changing rooms, towel and bathing costume hire, etc.), or with recreational facilities (beaches, chutes and diving boards). At the end of the nineteenth century the first swimming schools appeared and the number of establishments on the river banks multiplied to meet the nascent demand for outdoor sports. Along the banks of the Seine, the Marne and the Oise, Île-de-France soon had dozens of landscaped beaches where you could hire a boat, bathe and take part in entertaining activities: swimming races, fishing competitions and regattas. All these entertaining activities, along with the river cafés, were really like amusement parks.

In the inter-war years, bathing establishments included all sorts of nautical activities and came to be called "beaches", rivalling with the seaside resorts. They were still quite popular after World War II, but the advent of the indoor swimming pool led to their demise.

Riverside Dance Cafés

The story of the riverside dance café or "*guinguette*" is connected to the leisure time of the working class and their "Sundays down by the riverside". At the beginning of the twentieth century, people went to the river on Sundays and bank holidays for a variety of leisure pursuits, often related to the water: regattas, boating, swimming, diving competitions, rowing races, jousting or fishing. Emile Huet, in 1900, was full of praise: "*we hear droning and marching, like fireflies on a warm evening, tiny lights come and go on the water [. . .] and there are bursts of laughter. People dance and sing at the Eldorado whose two-storey pavilion lights up and fills with sound, and the moon rises and scatters drops of silver into the wakes of the boats*".

The guinguette should be placed in the more general context of the time: there was a wide variety of activities available to the public along the edges of the Marne River. During the summer season, nautical fêtes were organised in each town: in Joinville, it was the *fête des Ondines*, in Nogent, the *fêtes du viaduc*. These fêtes attracted thousands of visitors to the banks and on the water, to watch jousting, rowing races and flowered boat contests which animated the edges of the river at the time. The pleasures of the water were within reach for everyone: all sorts of boats could be hired.

Boating

After 1815, boats became the first autonomous leisure vehicles. Sailboats (*canots à voiles*) were imported from Normandy, hence the name of "*canotage*" for sailing… and the craft were gradually improved and diversified. Around 1840, popular boating included all those who frequented the "loose" guinguettes. On the Seine, some of the people out for a walk liked to play the sailor, dressing up and re-enacting naval battles. But there were also the serious adepts of boating as sport who organised competitions and regattas on the model of the English sport.

The rise in popularity of nautical sports in the first half of the nineteenth century in the area around Paris was connected to the development of the inland waterways. In fact, before it was channelled, the Seine had been navigable only 160 days per year. Starting from 1838, the channelling of the river, using the technique of mobile dams, made up for the lack of water in the summer months, while facilitating navigation. This infrastructure afforded flow depths that were adequate in all seasons for the growing activities of nautical sports and leisure boating.

At the end of the 19th and beginning of the twentieth century, people began to look for new spaces for leisure activities that were closer to nature. The "Canoe club" founded in 1904 set itself the goal of exploring rivers and streams in France to enable the growth of nautical tourism (Sirost and Femenias 2012). They were looking for undeveloped rivers for the pleasure of confronting the natural elements (white water and rocks) and to experience the *frisson* of adventure. The practice of canoeing grew at that time and was part of an imaginary wild nature. Subsequently, canoeing was nonetheless practiced in sites neglected by navigation and developed specifically for the sport, which had become competitive. Certain navigable rivers were therefore put to a new use.

Angling

Fishing with a line and hook has long been a livelihood activity. Numerous iconographic documents attest to the practice. With advent of the leisure society, this activity experienced extraordinary growth among urban dwellers looking for a break from their daily work routine, or trying to find a nostalgic lost world as the rural exodus emptied the countryside. Many fishing clubs were formed at the time and organised highly prized competitions. Moreover, in the years between the wars, this activity became a real social phenomenon arousing a huge amount of interest. In parallel, highly technical methods of fishing for sport and as a social activity (light rods, artificial flies, and reels), emerged with clubs such as the Fishing Club founded in 1908.

"Re-conquering" the Banks

For a few decades, the urban banks of major rivers and in particular port wastelands enjoyed a renewal of interest and generated some large-scale projects. The river had become an essential feature of the urban landscape. It had become, at least in the way it was talked about, an object of leisure and a symbol of nature. Under cover of rallying slogans of the type "Make the banks accessible" or "Put the river back at the heart of the town", the spaces abandoned by the decline in river transport or the delocalisation of industrial activities were refurbished to create areas for leisure activities.

In the towns, two major trends then emerged and sometimes interfered: first of all, promote the river heritage and give citizens back spaces of freedom and for recreational activities in relation to the waterways; secondly, redefine the waterways as "natural" ecological systems, or supposedly natural systems in the name of the preservation of biodiversity.

The abandoned hangars on the quays of the Garonne in Bordeaux, which were disused since the port closed in 1986, are going to be rehabilitated and reclassified. It is a question of renewing with the architectural heritage from the city's river navigation and port tradition by transforming these abandoned wastelands into busy cultural venues and retail units, in order to reconcile (or so they say...) Bordeaux citizens with the heritage of the river. Regarding the Rhône, the idea was to recover a "river that is lively and running". We are insistently reminded that Lyon owes its existence and prosperity to the river, and that by restoring it, it is the very nature of the town that is being rehabilitated. This touches on one of the founding myths and the river becomes the medium for a new urban identity. The river "deteriorated" by pollution and developments is contrasted with the conventional image of a time when man, who had few technological resources, lived in close contact with the river. And, following this anxiety-provoking rationale, if the river has lost its identity, people, for their part, have lost their soul.

> The emergence of new sensitivities and high social demand for public spaces and areas for leisure in cities, are making rivers a central preoccupation once more. This return to the river offered by free spaces and an open horizon seems in fact to correspond to a need for nature or rather for the idea we have construed of it. River cities are rediscovering their watersides, all the qualities of which they had denied, along with their strong potential for the renewal of the town's image. The numerous examples of revalorising the waterfronts undertaken all over the world, with the landscaping of spaces for leisure and cultural activities, retail, offices and accommodation, have snowballed, revealing to the city dwellers the pleasure of living near the water. (Lechner 2006)

Starting from the 1980s, the river regained a dominant place in the town (demonstrations and fetes on the banks, landscaping of parks, etc.). It is now endowed with new values drawing on essentially cultural contents (symbolic, recreational, and entertaining), for the well-being of the citizens. Water bodies are developed for their landscape and ecological values. Their banks, towpaths or quays, opened up to pedestrians or cyclists, become places for strolling.

Let us be realistic nonetheless, the rush towards the water is not entirely disinterested. The building sector benefits greatly and the prices of land and real estate have skyrocketed. Beyond the official discourse that glorifies the re-establishing of bonds between city dweller and river, some observers are less optimistic. If we refer to the past, the time of the riverside cabaret has gone. The guinguette has given way to a few restaurants where the price of the cheapest menu discourages the modest purse. The time for boating or even bathing in the river has also passed (bathing in the Marne was finally prohibited in 1970 for health and safety reasons). Nor are there any swimming pools at the water's edge any longer... You have to leave the urban centres behind to find a certain complicity with nature. And angling is no longer the favourite activity of the urban citizen... In terms of freshwater, nautical activities have most often migrated rather towards the reservoirs created for irrigation or for producing hydroelectric power, or even for flood protection. Sailing is practiced there, as in the sea, which still remains the principal centre of attraction.

France is one of the rare Western countries to have seen the appearance of a new type of use of the riverbanks, with the tethering of residential barges in cities, principally in Paris and Lyon.

Restoration?

In the 1980s, there was growing awareness of environmental issues and a new conception emerged of the waterway, which was no longer a physical object, but an ecosystem and a habitat. Its development therefore had to meet increasingly strict conditions in terms of ecosystem preservation. The new challenges are to restore the waterways as a living environment.

Restoring the waterways is an undertaking separate from the rewilding of the riverbanks. It is achieved through two main procedures: controlling pollution and restoring aquatic habitats. The control of pollution has undoubtedly been a success. Since the water agencies were set up in 1964, the measures taken have borne fruit, even though the situation is not yet fully satisfactory. As for the ecological restoration of the waterways, it has become a leitmotif on the part of managers and ecology movements. But beyond the general discourse that stigmatises the impact of human activities, the objectives remain vague.

In actual fact, the reasons behind most of the waterway restoration projects are aesthetic, economic or security related, and not the protection or restoration of biodiversity (Morandi and Piegay 2011). The situation becomes Kafkaesque when we look closer, for most of the projects fail to implement long-term monitoring, which would provide knowledge of whether or not goals were being reached. In a review of several hundred projects, Morandi and Piegay (2011) highlighted the lack of capitalisation and information exchanges on the success or failure of operations. In reality, there is little monitoring that could show whether or not the actions implemented had positive repercussions on the natural milieu. We, therefore, have the impression that what matters is the "announcement" effect!

Behind the slogans intended to rally citizen opinion around so-called restoration projects, what is really being promoted is a pleasant setting for living, in a space that has been rewilded, is healthy, attractive and above all safe. For we must not lose sight of the economic and industrial functions of the river, or of the fact that it is a constant threat because of the spates. Under cover of naturalness we are therefore seeing a staging of nature...

The story of the Bièvre in Paris, for instance, is very instructive (Le Roux 2010). First of all fitted with mills as early as the eleventh century, it was colonised in the fourteenth century by tanners and dyers, then by cobblers, launderers and weavers... and by the famous Gobelins Manufactory. At the beginning of the nineteenth century, the Bièvre became as polluted as a sewer, emitting such a terrible stench, that, for reasons of health, people wanted it removed during the works conducted by Baron Haussmann in the mid-nineteenth century. The Bièvre was therefore covered over and rerouted to the Paris sewers (Carré et al. 2011; Lestel and Carré 2017). To respond to the back-to-nature aspirations of certain citizens, large sums of money are now being spent to restore the course of the Bièvre, symbol of a river sacrificed in the name of the economy... The public has been mobilised around a falsely bucolic image of the river. Beyond the discourse on wilding and the conservation of biodiversity, in reality, what is at work is the creation of a landscape that is artificial but aesthetically pleasing, according to the norms of the moment, and secure, to prevent negative-impact events at all costs.

> It appears that what people call "nature" is eminently cultural. It does not coincide with the objective definitions given by the ecologists. While landscapes perceived as natural are deemed more aesthetically pleasing, this is because the landscape structures that characterise them correspond to a cultural ideal... These aesthetic preferences do not therefore seem, under any circumstances, to be related to good ecological status. In fact, naturalness is not necessarily compatible with the state of nature as it is apprehended by the ecologists. In this sense, naturalness, as perceived by the man in the street, is likely to be in contradiction with naturalness as apprehended by the ecologists in charge of defining good ecological status. (Tronchère-Cottet 2010)

Bellenger et al. (2015) drew our attention to the more or less underground role of local stakeholders. Alongside environmental restoration operations on the Seine estuary, conducted under the aegis of the government, a handful of nature lovers (hunters, fishermen, gatherers and naturalists) strive to make these processes of reconquering, restoration and rewilding coincide with the interests of the local population. They share a common passion, that of the garden estuary. Their interventions take the form of political pressure through play acting, for example on the subject of land planning schemes, but also by making active proposals in terms of rewilding (educational projects, heritage management, defence and development of uses, etc.).

River Heritage

Urban leisure activities developed within the largely manmade scope of the river landscapes inherited from developments undertaken to facilitate various uses made of the waterways… These are sometimes described as heritage.

Heritage: "*Any trace of the past that is tangible, intangible and symbolic, steeped in meaning for a community that appropriates it and endows itself with the means of collective transmission. It conjugates multiple facets, including those of eco-heritage, with the landscapes, the milieus, its flora and fauna but also with nautical practices*". (Le Sueur 2012)

Like rural heritage, river heritage, therefore, combines natural elements, built elements, structured landscapes, tools and a whole range of cultural elements related to the use and knowledge of the waterways.

The recognition of the heritage content of the rivers and their urban fronts has been a real driving force in the transformation of the perception and handling of the quays, banks, warehouses and industrial structures. The heritage value of a waterway then becomes a major stake for planning policies through conservation and development projects that are sometimes in competition.

The heritage approach is also a response to an identity-related motivation. The current heritage is our history, that of our ancestors who helped to build the setting for our home life. An approach that is sometimes idealised, or even invented. Thus, the idea of a Loire Valley identity based on a close link between the residents and the river (some people even speak of symbiosis) raises questions, since historical analysis shows that from as early as the eighteenth century the Loire was no longer indispensable to the agricultural economy. As for navigation, it ceased in the nineteenth century, and people living near the river, at least in the cities, kept their distance from the water which at that time was perceived as a threat due to disastrous flooding when the river was in spate.

The banks of the Seine in Paris are listed as a UNESCO world heritage site. From the Louvre to the Eiffel tower, or from Place de la Concorde to the Grand Palais and the Petit Palais, with architectural masterpieces such as Notre-Dame Cathedral and the Sainte-Chapelle, we can see how Paris has changed and watch its history from the Seine. Here, the river is not the main target but constitutes the medium of heritage.

Heritage policies must include the different representations and different economic challenges that interact in a territory. For some, it is a question of preserving a territorial identity and protecting it from damage related to opening up to the outside world. For others, on the contrary, heritage is a resource to be promoted as part of local development, in particular by opening up to tourism. And so the application to have the Loire Valley listed as a world heritage site was no doubt to showcase the historical significance of the region, but it is clear that the world heritage listing is also an excellent argument for promoting tourism. For behind the notion of heritage a new use of the waterways is emerging: tourism. The countryside is reinvested with respect to this new reference to legitimise a territorial development policy.

Folklore events such as gatherings of old sailing boats (Armada, Orléans) (Sirost et al. 2012) are another form of heritage approach. They are in fact driven by enthusiasts nostalgic for river cultures that are being lost and who wish to maintain the memory of the traditions, but they are also, and perhaps above all, based on economic considerations and the promotion of tourism in the regions.

The conservation of heritage comes into conflict with the principles of restoration of the waterways, and the dogma of ecological continuity that emerged from the Grenelle environment forum and carried by militant ecologists for whom the return to the wild is a priority goal. According to some managers, the levelling of dams would make it possible to re-establish the ecological and sediment continuity of the river, to meet the demands of good ecological status of the waterways as per the European directive on water. They argue that the vast majority of hydraulic facilities have lost their initial economic function, and that the high cost of maintenance, as well as their impact on the natural milieu, raise the question of what is to become of them. This, of course, refers to small structures such as mills, some of which are undoubtedly in a state of abandonment. But this approach makes little of the residential river landscape which has been built up over time, and the attachment of the residents to their mills and the associated reservoirs which also serve various ecological purposes.

Conclusions

Our river systems have been greatly altered down through the centuries to meet the needs of productive or defensive uses some of which are now receding or are no longer current. Many of these developments have since been "recovered" for other activities, in particular recreational. There are numerous examples: tow paths for hiking, dams and reservoirs for sailing, millstreams for canoeing or kayaking, canals for river tourism, etc. In this context, cities have undertaken to transform their abandoned river infrastructure into spaces for recreational and entertaining activities. Under the pretext of re-establishing the link between river and citizen, which is partly true, large-scale real estate operations have been arranged that are of benefit to the city's economy. In this type of operation, which consists of staging nature using our current criteria of aesthetically pleasing and safe, in reality little concern is given to wildness, while the impression is given that a natural river is being recreated. The banks are still covered in stone and concrete, and there is no question of allowing the river to flow freely once again. . .

S. Bonin (2007) invites us to reconsider with a critical eye the discourse presented by the towns that "breathe new life" into their river, "reconcile with..", or "set out to re-conquer. . .". She tells us clearly that there is no link between this dynamic and a rise in environmental concerns for hydro-ecosystems as a living habitat in the town. "*'Natural' banks, islands, riverside woods, quality water, like the uses we see supporting projects in ecological engineering (environmental education, local leisure pursuits such as fishing or bathing) are not the subjects of this re-conquest*". And she supports her words by underlining that this type of concern disappears very early on in the realisation of two urban projects studied: the islands of the Rhône downstream of Lyon, and l'île de Nantes, on the Loire.

References

Barrier P (1990) La mémoire des fleuves de France. C. de Bartillat, Etrepilly

Beauchêne S (2006) Les fleuves dans le processus de métropolisation de l'agglomération Lyonnaise. Editions MdfR, Givors

Bellenger MC, Machemehl C, Sirost O (2015) Les passeurs de nature face à la question de renaturation des fleuves. Chasseurs, cueilleurs, pêcheurs de la vallée de la Seine. ISI Rivers. https://www.graie.org/ISRivers/docs/papers/2P21-49588BEL.pdf

Berthier K. (2009) D'une rive à l'autre. Service Etudes et Programmation, Direction des Espaces Verts et du Paysage Conseil général du Val-de-Marne. http://docplayer.fr/11298555-Service-etudes-et-programmation-direction-des-espaces-verts-et-du-paysage-conseil-general-du-val-de-marne-karine-berthier.html

Bonin S (2007) Fleuves en ville : enjeux écologiques et projets urbains. Strates [En ligne], 13. http://strates.revues.org/5963

Bravard JP (2004) Villes de réservoirs sur le Yangzi et sur le Rhône : niveaux fluviaux et gestion des berges à Chongqing et Lyon. Géocarrefour 79(1):49–62

Carré C, et al. (2011) Les petites rivières urbaines d'Ile de France. Découvrir leur fonctionnement pour comprendre les enjeux autour de leur gestion et de la reconquête de la qualité de l'eau. PIREN Seine, fascicule 11. http://www.sisyphe.upmc.fr/piren_drupal6/?q=webfm_send/1008

Guillerme A (1990a) Eaux vives et eaux mortes entre Moyen Age et Renaissance. in « Le grand ivre de l'Eau », sous la direction de Jean Aubouin. La Manufacture, Paris

Guillerme A (1990b) Le testament de la Seine. Géocarrefour 65-4:240–250

Huet E (1900) Promenades pittoresques dans le Loiret (Châteaux – Monuments – Paysages), 325 p

Le Roux T (2010) Une rivière industrielle avant l'industrialisation : la Bièvre et le fardeau de la prédestination, 1670–1830. Géocarrefour 85(3):193–207

Le Sueur B (2012) Navigation intérieure : histoire de la batellerie de la préhistoire à demain. Douarnenez, Chasse-Marée, 240p

Lechner G (2006) Le fleuve dans la ville. La valorisation des berges en milieu urbain Note de synthèse, Direction générale de l'urbanisme, de l'habitat et de la construction, centre documentaire de l'urbanisme, octobre 2006

Lestel L, Carré C (2017) Les rivières urbaines et leur pollution. Quae, Paris

Lévêque C (2016) Quelles rivières pour demain? Quae, Paris

Lévêque C (2019) La mémoire des fleuves et des rivières. L'histoire des relations entre les hommes et les cours d'eau à travers les siècles. Ulmer, Paris

Miquel P (1994) Histoire des canaux, fleuves et rivières de France. Éditions no 1, Paris, 317p

Morandi B, Piegay H (2011) Les restaurations de rivières sur Internet: premier bilan. Natures Sciences Sociétés 19:224–235

Porhel JL (2012) Histoire des incessants travaux menés par la ville de Tours pour lutter contre les inondations. Mémoires de l'Académie des Sciences, Arts et Belles-Lettres de Touraine 25(2012):153–184

Sirost O, Femenias D (2012) Les usages récréatifs de l'estuaire de la Seine. GIP Seine Aval, Rouen

Sirost O, Melin H, Lecoeur M, Bouillon D (2012) Projet Enfants Du Fleuve: Paysages vécus et paysages perçus de l'estuaire de la Seine. Projet Seine-Aval 4, 223 p. http://www.seine-aval.fr/projet/enfants-du-fleuve/

Tronchère-Cottet M (2010) La perception des bras morts fluviaux : le paysage, un médiateur pour l'action dans le cadre de l'ingénierie de la restauration. Approche conceptuelle et méthodologique appliquée aux cas de l'Ain et du Rhône. Thèse université Lyon III, 361 p

With Rivers to the Sea: Ecological Restoration of Rivers and Estuaries and Nature-Based Activities

Jean-Paul Ducrotoy

Summary

Estuaries are coveted territories. However, most of the time they are threatened by human activities, which are meant to make the best use of them. They can be so degraded that the restoration of their original ecological functions is necessary. To understand what measures are necessary to restore some ecologically weakened functions in a river, its estuary and watershed, some examples are drawn from the Interreg Tide program, which took place from 2010 to 2013. In this chapter, an ecosystem approach has been adopted to propose improvements in tidal estuaries of the English Channel and the North Sea.

This chapter further considers how the public perceives such renatured environments in estuaries where motives for reconquering damaged estuarine habitats were quite different In the Elbe sediments were managed for navigation to the port of Hamburg (Germany). In the Scheldt flooding was a problem in relation to keeping access to the port of Antwerp (Belgium). In the Crouch sediments extracted from London underground for constructing a fast metro were used to create a natural park at Wallasea (England). In the Seine compensation measures were taken due to the development of port facilities at le Havre (France). The use of restored or recreated habitats for recreational activities was neglected. These case studies show that, despite the existence of few success stories, the public often misunderstood the need for re-establishing ecological functions in ecosystems, which had much more to offer than cycle paths and sailing facilities. Paradoxically, humans are rarely considered as an integral part of the ecosystem and sociologists have promoted the term "socio-ecosystem" to reinforce the fact that human societies do belong to their environment.

The conclusion emphasizes the need for an anthropological approach to back and sustain ecological studies.

Keywords

Estuary · Habitat restoration · Ecological functions · Recreational activities · English Channel · North Sea

Introduction: Ecological Restoration Versus "Nature"

When rivers reach the sea, their ecological characteristics get richer due to the formation of a particularly productive exchange zone between the watershed, groundwater, the atmosphere and the marine environment. The mouth of a river can be considered an estuary when it has a free connection with the open sea, and within which seawater is measurably diluted with freshwater derived from land drainage. In tidal estuaries, currents create ecological gradients from the open sea up to the head of the estuary. When the marine hydrodynamics are strong due to a large tidal range, strong swell and strong littoral currents, vast intertidal zones are formed. Longitudinally, an estuary extends to the zone of influence of the tides. The river brings in few coarse materials but carries particles and soluble substances. Transversally, an estuarine system, as an ecological entity integrates adjacent areas, usually wetlands, and, depending on the context, the minor and major beds of a river. The landscape offers a diverse but homogeneous outlook.

Biogeochemical cycles and ecological processes interact and are interconnected in a complex ecosystem enabling humans to derive many and diverse ecosystem services. This explains why estuaries are sought after by humans, both for ecological and socio-economic reasons. Based on their geomorphological setting, the value attributed to estuaries by human societies arise from their astonishing biological productivity as well as on the possibility of

J.-P. Ducrotoy (✉)
Institute of Estuarine and Coastal Studies, The University of Hull, Hull, UK
e-mail: Jean-Paul.Ducrotoy@hull.ac.uk

C. Machemehl et al. (eds.), *Reclaiming and Rewilding River Cities for Outdoor Recreation*, Estuaries of the World,
https://doi.org/10.1007/978-3-030-48709-6_3

establishing ports favourable to trade and industry. However, the variety and intensity of anthropogenic pressures might create ecological perturbations which range from local to global (Crossland et al. 2005). Today's challenge is to reconcile unregulated coastal management and intense exploitation of living resources with a galloping drive for economic gains (Vinebrooke et al. 2004). Among the known impacts, habitat loss or degradation are considered the most detrimental not only to human activities but also to the overall functioning of the ecosystem.

Recently, a real need for "renaturing" overexploited sites has been expressed in European estuaries, with facilities required for outdoor recreational activities ranging from waterfowl hunting to water sports and bird watching. That is why in recent years, post-work compensation (mitigation) has been used for restoring degraded habitats or recreating lost habitats in order to maintain or rehabilitate ecological functionalities threatened or lost.

The loss can be permanent, the wetland disappearing purely and simply, converted into an industrial port or agricultural zone. The construction of polders has long been the response in Europe for a need for land. Thus, in virtually all tidal estuaries in the North-West of Europe, vast areas of salt marshes have been dammed and transformed into industrial zones, or converted into pasture and other agricultural lands. In the Bay of Somme (France), about 280 km^2 have been conquered over the sea during the last four centuries, on an intertidal zone that covered more than 350 km^2 in the sixteenth century (Ducrotoy 2017).

Degradation may also be temporary or localized, most often as a result of pollution in the form of contaminated discharge. These two types of degradation can only be remedied by decontamination, restoration or recreation of lost biotopes (Ducrotoy 2010). It will be seen in this chapter that ecological restoration covers those strategies which lead to the "recovery" of an estuarine ecosystem that has been degraded, damaged or destroyed (Palmer et al. 2016). Such an approach is relatively recent and is based on ecological engineering, which has allowed the recent development of a new type of approach leading to re-estuarisation or de-polderisation actions.

The origin of this movement has its roots in the monastic life of the Middle Ages and in the romantic movement of the nineteenth century, founded on a spiritual (sometimes mystical) approach to nature. It has been concretized and made tangible thanks to the scientific revolution which favoured the emergence of ecological science. Over the past few decades, many have realized the value of the scientific approach, not only as an essential element of nature conservation, but also as a framework for a compromise between humans and their planet in a relationship that has become extremely conflictual.

The striking progress of biology and the expansion of its field of investigation led to a shift of focus where nature became a priority and was worthy of the attention of engineers. The first experiments to try to recreate natural ecosystems were carried out at the beginning of the twentieth century by pioneers who first considered an ecosystem as a whole (Leopold 1949). The reasons for this revolution are multiple and complex: a mixture of scientific but also historical curiosity, caused by an aesthetic interest and a certain nostalgia. From a scientific point of view, respect for the old and the established cannot be avoided with the idea that a "climax" ideally combines a privileged ecological assemblage of organisms, endowed with distinctive qualities of stability, harmony and a capacity for self-organization. Such concern for what will later be called "restoration of natural habitats" will remain unexpressed until the end of the twentieth century. The idea of recreating a complete ecosystem or even an ecological community or landscape, with all its attributes and processes, had long been considered utopian (Jordan and Lubick 2011). Most often, such improvements result in the opening of hitherto condemned spaces, which were often unhealthy or, at least, damaged and degraded, which the public do not necessarily regard as prime locations for the development for leisure and tourism. However, the leisure industry and the practice of sports by more and more people (generally equipped with sophisticated gears) has seized the opportunity in the years 1970s, in establishing now popular leisure centres and nature trails in estuarine restored areas. But how can these so-called nature-based activities be integrated into the development plans of regenerated habitats?

This chapter considers how the public perceives such re-natured environments in several North-West-European estuaries where motives for reconquering damaged estuarine habitats were quite different. It shows that despite the existence of few success stories, the general public often misunderstands the need for re-establishing ecological functions in ecosystems that have much more to offer than cycle paths and sailing facilities.

Estuaries: Coveted Territories

To understand what measures are necessary to restore some ecologically weakened functions in a river, its estuary and watershed, some examples are drawn from the Interreg Tide program[1] which took place from 2010 to 2013. In this chapter, the same approach has been adopted later for additional

[1] The "Tide—Tidal River Development" research program was implemented between 2010 and 2014 with co-funding from port authorities and the European Union. Confronted with the protection of natural environments (protected by European directives) and the strong economic needs expressed by seaports, Tide brought together European experts from universities, ports, waterways administrations, etc. Management practices in the Elbe (De), Weser (De), Scheldt (BE/NL) and Humber (UK) have been compared to find solutions for sustainable development of ecosystems.

tidal estuaries of the English Channel and the North Sea. These coastal seas form a continental system of the north-eastern Atlantic, one of the few large marine ecosystems formed by the recent flooding of a landmass that occurred 20,000 years ago (Ducrotoy et al. 2000). This makes it a young ecosystem, and therefore in a state of rapid ecological change, still today. The southern part, which includes the Channel, is shallow and presents conditions with strong tidal currents where the estuaries selected for this presentation developed: the Somme and the Seine (France), the Crouch in England, the Scheldt in Belgium and the Netherlands, the Elbe in Germany. All of these estuaries have macro- to mega-tidal characteristics and have been partially filled by marine sediments because of the Flandrian marine transgression which began about 3000 years ago. As a consequence of the resulting plugging of the internal zones, there is presently a progradation of sedimentation towards the open sea. This has favoured the emergence of flood plains, which must be considered as an integral part of the estuarine complex, even if they are largely occupied by humans. The geomorphological mechanisms involved have facilitated the action of local populations which have isolated many territories by the construction of polders and ports. Thus, large areas of marine habitats have become terrestrial. With increasing industrialization and the development of maritime transport activities, the trend accelerated throughout the twentieth century, resulting in considerable loss of intertidal areas and sharp changes in local geomorphology. Although all the estuaries studied in this paper fit into this general scheme, it is of note that the restoration of damaged habitats and ecological functions in each of these estuaries has benefited from quite different management methods. In the estuary of the Seine (Normandy, France) 90% of intertidal zones were lost, whereas in the estuary of the Crouch (Essex, England), the Royal Society for the Protection of Birds is preparing to open an immense intertidal natural reserve, reconquered on agricultural land. For each example an attempt is made to analyze consequences in three stages: the situation of ecological restoration actions, the level of urban development of port cities and the practice of "nature" recreation (e.g. water bodies, fishing areas, bird watching areas, playgrounds, cycling routes, walking trails...) considered to be ecosystem services to society.

Why Restore Degraded Estuarian Habitats?

In its simplest form, ecosystem restoration can be defined as a set of initiatives taken to improve an ecosystem, enabling it to attain a better state, whatever it may be (Livingston 2006). This definition does not imply a return to a previous state, but only concerns the restoration of certain ecological functions. A definition, considered more orthodox, refers to the processes used to restore habitats, or an ecosystem considered to be viable both in its structure and in its functioning, according to the degradations inflicted by human activities (Elliott et al. 2006). In summary, restoring an ecosystem will reinstate the functions lost by this ecosystem, or replace some functions which are considered superfluous. In a radical way, it will be possible to decide on the replacement of the affected ecosystem (or certain habitats) by another ecosystem, which will fulfil the intended functions. Habitat recreation can thus lead to opposite results when considering estuarine systems. In the case of restoration, a system may have succeeded in maintaining or even gaining estuarine characteristics; in the other case, the estuary will no longer exist as such.

It should not be forgotten that the functioning of any ecosystem is based on physical, chemical and biological processes. They enable humans to derive benefits from the ecosystem services which flow from them, such as the possibility of outdoor recreation. For estuaries, accessibility to navigation is a rarely questioned service, much more appreciated than the regulating effect of vital continental flows towards the sea and the biodiversity they shelter. The Tide program showed that seven of the ten services with the highest demand were linked to hydro-geomorphological regulation and two were directly related to the nautical or industrial use of water. The provision of these ecosystem services varied spatially and temporally (e.g. seasonal variability) between habitats and between different systems. It is therefore in response to the need to maintain or restore these services that most of restoration operations were undertaken. The use of restored or recreated habitats for recreational activities is often neglected. Paradoxically, humans are rarely considered as an integral part of the ecosystem and sociologists have promoted the term "socio-ecosystem" to reinforce the fact that human societies do belong to their environment.

How Does the Public Perceive Restoration?

Putting an ecological restoration project in a scientific perspective implies the application of the fundamental principles of ecology. However, due to the popularity and misunderstanding of certain concepts such as biodiversity or biological productivity, the approach might be biased. Systematic use of consecrated concepts might go adrift and obscure the real stakes of the socio-economic development of the estuary (Ducrotoy and Yanagi 2008). For example, for the general public and politicians, biodiversity is the basis of robust and productive ecosystems, which has never been scientifically demonstrated. Recently, Elliott and Quintino (2007) highlighted the paradox of estuarine quality, where a rudimentary specific richness supports high production and ensures the stability of a highly dynamic system (Holling 1978; Peterson et al. 2010). To understand the serious political issues underlying the implementation of often complex estuarine restoration projects, a few examples follow to illustrate how, nevertheless, a (sometimes tenuous) link can be recreated between the inhabitants and the environment. We can judge it through the study of outdoor recreation, bringing in a scientist's perspective on the relationship between "nature", recreation and communities and exploring the main sociological issues of an "ecological" restoration policy.

The Elbe: Managing Sediments for Navigation to the Port of Hamburg (Germany)

The first case study is the estuary of the Elbe (Germany). To improve navigation and reverse the steady increase in activities devoted to dredging the estuary, the Port of Hamburg and the federal service of inland waterways (WSV) have developed a concept to mitigate tidal energy responsible for bringing into the estuary increasing amounts of sediment. The first goal was to reduce dredging activities, while promoting conservation of existing natural habitats, to enhance biodiversity and to promote the activities of "nature" such as fishing and outdoor recreation. In Germany, the Länder and the federal government oversee the implementation of European directives. To preserve the habitats included within the Natura 2000 network, the federal states and the WSV developed an integrated approach for the development of a management plan for the estuary. It was intended to integrate economic and ecological requirements through a set of measures considering many constraints such as navigation channel maintenance, agriculture and recreation. The plan was not legally binding, it was based on the voluntary commitment of stakeholders. Regarding the dissipation of energy of the tide in the mouth, "natural" river engineering controls were put forward. The main management problems stemmed from conflicts marked between the navigation of larger vessels and maintaining the areas protected under the Natura 2000 network (Fig. 1).

A sociological survey (Ratter and Weig 2012) was conducted in 2012–2013 by Tide to establish how local populations perceived proposed strategies. The investigation revealed weaknesses in the knowledge of the public on both the natural characteristics of the estuary and the necessity to dredge it regularly. The idea of nature remained vague although it was possible to connect (often contrasting) opinions to specific social groups. Some saw nature as virgin, where humans had never intervened, so non-existent in Germany. Others had a vision more down to earth and considered natural any territory that had some natural elements such as trees and birds. Regarding the estuary, everyone agreed that it was an indispensable communication axis but that they did not hesitate to visit regularly for leisure.

Fig. 1 Haseldorf Nature Reserve, the Elbe estuary and Hamburg industrial estate in the background

However, responses regarding the type of outdoor activity often lacked precision. The natural element which appealed to most respondents was the tide but there was little about the flora and fauna. For example, the saffron dropwort *Enanthe conioides*, a protected endemic plant, or the feigned shad *Alosa fallax*, an estuarine fish, were not particularly popular with the public and mostly unknown. The "Natura 2000" appellation was unfamiliar to residents who often overestimated the space devoted to conservation. As for recent or ongoing developments, the public remained rather insensitive to restoration operations.

The Scheldt: Preventing Flooding While Keeping Access to the Port of Antwerp (Belgium)

In the example of the estuary of the Scheldt arises the need for integrated management of water to achieve security goals for populations in the face of potential floods, while improving accessibility to the port of Antwerp and the need for community-based natural recreation in a highly urbanized area. Such an approach rested on an ambitious practice of estuarine ecosystem restoration and rehabilitation of ecological processes pertaining to the original systems (such as the recycling of nutrients or other hydrological processes). Based on an analysis and an assessment of current and future problems, both in the valley and in the estuary, a conceptual model of rehabilitation with a long-term vision was developed. Landscaping was inspired by a trade-off between safety and ecology by the creation of flood control areas (FCA), placed under the influence of an artificially controlled tide (Controlled Reduced Tide, CRT) (Temmerman et al. 2013). To do this, selected polders were reconnected to the river (de-polderisation) by a system of regulators allowing a controlled flood, according to the state of the tide in the river (Fig. 2). To build the system of sluice feeding the FCA, mathematical models were used (Temmerman et al. 2003a; Ysebaert et al. 2005). Recently processed sites could perform important ecological functions with effects on water aeration, nitrification, denitrification, sedimentation and primary production (Temmerman et al. 2003b). The ecological redevelopment of polder areas thanks to CRT systems was case specific (with regard to elevation and bathymetry in particular), depending on, for example the morphology of the sector, the design of the lock or the quality of the water of the river. It should be noted that the quality of the water in the estuary would be improved indirectly because sedimentation could be reduced thanks to the development of FCAs (Eertman et al. 2002) (Fig. 3).

Fig. 2 Construction of a floodgate to control the tide (Controlled Reduced Tide, CRT) in an area in the process of de-polderisation in the Scheldt (Belgium)

Fig. 3 Bay of Somme (France). (**a**) Artificial ponds. (**b**) Hunting pond and hide with plastic lure birds to attract migrating birds

The Somme: Hunting and Tourism in the Face of Silting and Hypothetical De-polderisation (France)

Nested between European industrial hotspots, the Baie de Somme is known as a "wild" nature monument squeezed between urbanized poles in France, Belgium, England and Germany, with a reputation for authenticity and isolation. It attracts many tourists seeking to relax and consume heritage. The importance of the site is highlighted by both financial and equity investments for protection at national and European levels (nature reserve, orders of biotopes, important bird area, NATURA 2000, Conservatory of coastal areas and lakeside shores, area of flora and fauna habitat and of geological and natural interest, visitor centres, etc.). Traditional activities such as fishing have contributed to the territorial identity of the site, but hunting has precedence. Historically, its practice is linked both to the existence of large hunting areas and concessions granted by the Maritime Affairs to individuals who request digging an artificial pond and the benefit of a floating hide. During much of the twentieth century, these spaces have been voluntarily kept in a pseudo-wild state with ostentatious leisure purposes. The hunting hobby is still popular today. It shows that local identities are struggling to evolve, alongside (and often in conflict with) others which emerge such as bird-watching and walking. Hunters argue to be essential territorial actors, claiming that hunting waterfowl is vital to the local economy. Spaces voluntarily maintained "huntable" (scores of artificial ponds, each equipped with a metallic hide), are so-called "green infrastructures" which, unexpectedly, have contributed to the reputation of wilderness of the area. Initially, they have attracted supplementary tourist activities and have provided the basis for the conversion of the bay, just like other French estuarine sites with a natural and wild character. This notion of wilderness was based on an odd idea of biodiversity far from scientific reality, based on the collective memory of hunting societies and attached traditions. This way of considering natural resources reflects the behaviour of a social group with deep far-right political traditions.[2] Such conceptions have long conflicted with a rational management of the environment and were responsible for blocking current restoration projects that have been in the pipeline for more than a decade (Ducrotoy and Kalaora 2009).

The Crouch: Use of Sediments Extracted from London Underground for Constructing a Fast Metro to Create a Natural Park at Wallasea (England)

The Wild Coast project run by the Royal Society for the Protection of Birds (RSPB) in Wallasea Island had for purpose to get rid of huge quantities of sediment produced by the digging of railway tunnels under the city of London for the Cross-Rail project. 7.5 million m^3 of inert embankments had to transit by sea from London and were distributed on nearly 800 ha of polders in a space between the estuaries of the Crouch and the Roach, North of the Thames (Fig. 4).

Four environmental objectives were associated with (1) recreating intertidal habitat to compensate for intertidal habitats losses the last century and thus restoring living species in support of the Government of the United Kingdom's Biodiversity Action Plan. The removal of seawalls will be enough to raise the bathymetric level of the site to create mudflats, salt marshes, brackish lagoons and halophile pastures; (2) avoiding flood risk related to the recent construction of a poorly conceived dam that should be open by six breaches of 100 m wide; (3) creating a recreation area, centred on the enjoyment of natural features, both for birdwatching and walking; and (4) demonstrating that it is possible for society to adapt to climate change, particularly in view of sea level rise in estuarine and coastal sites.

Eventually, in 2019, the site may receive several thousand visitors. Two car parks with restrooms, restaurants, snack bars and Park House are planned as well as 15 km of access roads, which will be built to connect the Park to main roads. The Wallasea Island Wild Coast project (https://www.rspb.org.uk/reserves-and-events/find-a-reserve/reserves-a-z/reserves-by-name/w/wallaseaisland/) is probably the most ambitious in the world so far in what is called the restoration of coastal wetlands with a clear scope for nature-based activities.

The Seine: Compensation in the Face of Development of Port Facilities at Le Havre (France)

The history of the ecological restoration of the Seine estuary began in the 1990s with the construction of "Port 2000", a large extension of the port of le Havre. After the construction

[2] Hunting, Fishing, Nature, Tradition (French: Chasse, pêche, nature et traditions—CPNT) is an agrarianist French political party established in 1985, which aims to defend traditional values of rural France.

Fig. 4 The Wallasea Island Wild Coast project area in the Crouch estuary (England)

of the new facility, compensatory measures were decided to respond to the protest of local associations (non-governmental organizations, NGOs) in order to preserve or recover some ecological functions of the estuary of the Seine (Hamm et al. 2004; Dauvin et al. 2006). Finally, the French national authorities decided to ratify the extension project of the port of le Havre in September 1998. Later, they recognized the need to find a balance between the objectives of economic development of the Port 2000 project and the protection of aquatic habitats by a management program specific to the estuary of the Seine (Ducrotoy and Dauvin 2008).

These measures include (1) the opening of seawalls along the main navigation channel; (2) the enlargement of a breach to allow the water to flow more easily in the direction of the new upstream opening; (3) the construction of a submersible dike in order to facilitate the flow of tidal water; (4) the extension of an existing dike; (5) dredging more than five million tons of sediments; and (6) the construction of artificial resting areas for birds on the shore, as well as an artificial island in the estuary (Dauvin et al. 2006).

Following execution of this work, a sociological study was conducted by Sirost et al. (2012) to verify if the populations concerned were aware of the compensatory actions taken after construction of Port 2000. Even though NGOs initiated the process, 11% only of those interviewed knew at least one of the compensatory measures. The resting areas for birds were among measures better known to the public as they were the most publicized by television and newspapers (Fig. 5).

Outdoor Recreation and Ecological Restoration

Landscape Approach

To understand how estuarine restored habitats fit in an urbanized and industrial universe or in a natural environment in full motion, we must put them in context of the surrounding landscape. Behind conservation, draws a landscape policy, which gives the territory a frozen vision. Nature is consumed in small doses without individual commitment.

In the Scheldt, the ecological vision resulted in plans and concrete implementations which attempt to combine the security measures against the floods with the rehabilitation of the tidal river. All of the planning processes resulted in the implementation of complex systems of regulation of tidal levels, gigantic metal and concrete gates regulating the polder flooding. In selected polders, entire domains have been rehabilitated and the fight against floods has become a useful framework for habitat restoration. Providing space for surplus water has turned into an opportunity for recreating

Fig. 5 Mitigation measures in the Seine estuary after the construction of Port 2000 in le Havre (France). (**a**) Artificial island for birds. (**b**) Construction of a sluice gate for de-poldering an area for birds

habitats associated with the estuary, potentially attractive for walkers, birdwatchers, etc. (Maris et al. 2007). The risk is that the landscape might become compartmentalized with leisure activities reduced to walking and cycling along old dikes, without being able to enter and experience the newly created salt marshes. Walkers and cyclists would seem to navigate among natural scenographies without having any direct experience of nature. Restoring habitats would thus be an isolated gesture resulting in a "patchwork" of mini-reserves used for passive leisure activities.

Sociological Approach

For Sirost (2009), leisure and outdoor activities, like any other bodily practices oriented towards nature such as camping, consist in a quest of clean air in response to urban life, and in an effort to escape away from the modern values of Western society. Among the range of practices that contributed to the territorial identity of many French estuaries, we mentioned hunting in the Baie de Somme. The practice of this hobby still today, shows that mentalities are struggling to evolve, alongside (and often in conflict) with a new approach to nature. Hunters claim to be territorial actors, saying that hunting waterfowl is at the basis of a traditional economy that should remain unaffected by any other developments. Spaces voluntarily modified for this activity have lost their natural characteristics but are reputed to be "green resources" (Fig. 4). This way of looking at biodiversity reflects the strong political stand of a restricted social group. They feel holding a way of life of which they are the only guarantors, out of any broader context. Despite substantial financial investments made by the General Council of the Somme for developing green tourism, part of the residents feels disenfranchised. Authorities and locals do not seem to speak the same language. Decisions taken by the authorities are seen as short term and, anyway, inconsistent. At the present time, there is a lack of trust which is conveyed by small local associations, whose survival is often based on a handful or even a single individual. Authorities respond to these concerns by organizing "briefing meetings" often interpreted as propaganda. More information on the strengths of the region could encourage greater connectivity and a cohesive community spirit.

In the Länder of Hamburg and Schleswig Holstein, local authorities and the public do not seem to speak the same language either. The local population, although culturally different from what we have described in Picardy, has little involvement in the planning and management process because, as in the Baie de Somme, residents feel insufficiently qualified. Decisions taken by the authorities are perceived as short-term, based on the interests of certain social groups. There is a lack of confidence, and this hinders cooperation with decision makers (Ratter and Weig 2012).

Anthropological Approach

In Essex, the Crouch and the Roach estuaries seem to be destined to become a kind of "Nature-Land", comparable to an amusement park where visitors come to consume their share of nature without truly getting any living experience. As in the Scheldt, a psychological artificialization of the landscape is to be feared in the name of hobbies which make actors of this pseudo-adventure passive consumers of nature. Ingold (2011) opposed the "residential" perspective ("dwelling perspective") to the "constructive" perspective ("perspective building"). The first addresses the immersion of the body-person in the living environment while the second suggests that people build their world, in all conscience, even before they can act on it. As a scientist, I feel called into question but also energized when Ingold asserts that a real cultural ecology must transcend the pragmatic conditions of adaptation to the environment to focus on the symbolic significance through which humans perceive their environment. Any approach to outdoor activities must take into account not only physical tools but also the knowledge and skills required to use them. It must be inspired by many things in everyday life, like walking, orienting, cycling, using a field book, i.e. a whole set of practices and human behaviours, which are intrinsically related to the environment in which they take place. It is probably, here, one of the keys to restore working landscapes significant to human societies.

Conclusions

Wetlands and estuarine aquatic environments are at risk, even when they are managed by good willing administrations. Intertidal areas are the most vulnerable because of the contemporary need for space to develop ports, industry and agriculture. Nearby cities also expand in or around estuaries. The need to restore degraded or destroyed estuarine habitats must be acknowledged and must be the subject of a genuine accord in the context of an ecosystem-based and interdisciplinary approach. Recent studies in the estuary of the Seine (Dauvin et al. 2006; Ducrotoy and Dauvin 2008) showed that essential ecological features were still present, but that they were more or less failing. Most often, the priority was to restore weakened functional links between the floodplain and the river, leading to an improvement of hydrodynamical and biological processes. However, examples in this chapter illustrate the lack of integration of restoration projects, which were seldom coordinated under an overall (holistic) approach. A long-term ecological vision, based on robust scientific knowledge and historical data, is recommended to build scenarios for the promotion of local human activities in harmony with the changing conditions of the estuary.

In all case studies presented in this chapter, it appears that outdoor activities were superadded to projects intended primarily to combat flooding or pollution, to dispose of large quantities of spoil or proposed as mitigation for loss of authentic estuarine habitats. The challenge for planners remained to develop industrial and port activities where humans could interact (and coexist) with natural processes. Thriving and dynamic ports must incorporate the natural world in their thinking.

The lack of integration of open-air activities reflects a psychological artificialization of the landscape in the name of nature. They make actors of this pseudo-adventure passive. Too often, restoration proponents think that simply recreating a "perfect" habitat will make it a place of life for local residents and residents of nearby towns. The temporality of the landscapes, the topology of the environment and the changeability of the weather can hamper their efforts. According to Ingold (2011), a real ecological culture must transcend the pragmatic conditions of adaptation to the environment to focus on symbolic significance levels through which humans perceive their environment. Vision, hearing and human movement determine how people create a "niche" in the world. The approach to outdoor activities must consider not only physical tools but also the knowledge and skills required to use them. For this, it must be inspired by a whole set of practices and human behaviours that are intrinsically related to the environment in which they take place, and not just sport technics similar to those used in indoor exercise.

In several case studies presented in this chapter, one can note an absence of consultation with local communities. It represents a major stumbling block in the implementation of policies of protection and valorisation of nature. Even in cases where there was a consensus on the need to rationally manage and retain original ecological features, the development of management policies could not dispense of phases of negotiation included in the project to meet the sociocultural requirements of the local population. We must remain aware that policies underlying any negotiations rely on terms of power-sharing, which shape the role and interests of stakeholders (Baker and Eckerberg 2013). A good understanding of adaptive management strategies is necessary to establish a dialogue between managers and other stakeholders, including on technical questions of management. However, the movement towards the professionalization and institutionalization of participation observed in recent years does not go without giving rise to any questioning and criticism. It can be focused against the risk of dilution of the decision, or even controlled generalization of the defiance which would promote participation. Sometimes it can even be directed against a technocracy of participation more favourable to the development of procedures than to the free development of criticism. This usually leads to the depoliticization of the issues which would, under more favourable circumstances, have led to a successful and beneficial participative democracy.

For the scientific expert, the landscape consists of plant and animal communities that are disseminated according to environmental conditions and which interact in a given geographical unit: the ecosystem. According to this definition, the human dimension is integrated de facto. Like other beings, humans adjust and try to adapt to their environment. Human activities are the result of political choices made by humans following models varying according to cultural backgrounds. Today, the scale of change imposed on nature multiplies. Frantic development, unsustainable exploitation of resources, and the commodification of living and mineral resources ruin landscapes, disconnecting users from their environment and threatening their sense of belonging to their place of living (sense of place). Paradoxically, restoration operations are often met with bewilderment or indifference from the inhabitants of the region. Some heritage fanatics can even react in a counterproductive way which rejects virtually all change in existing landscapes. Cannavo (2007) identifies this type of conflict between development and conservation as a major factor in a growing malaise, typical of urban and peri-urban environments. Cannavò offers practical and theoretical alternative solutions to this impasse, by offering an approach that encompasses both change and stability and unifies democratic and environmental values: the concept of a "functional landscape" or working landscape. He argues that the landscape is not a passive object, but the product of human actions on the physical and conceptual environment, organized in a sustainable and coherent system. Yet we should not fall into a romance, thinking that past situations were ideal. In fact, the current situation is not so radically different from past situations; it would be risky to believe that all philosophies passed, including any rationalist practice (as well as the use of technology), are no longer adequate. These are tools that we cannot throw away deliberately. H. Jonas (2000) argued that the boundary between State and Nature had been abolished. He goes on to declare that:

> The natural has been swallowed up by the artificial sphere.

Such a statement can only lead to inaction. It would lead to technophobia.

The best framework for the incorporation of conservation in development strategies would be by democratic governance at the regional level (decentralized but not independent). We are still very far in Europe from a governance based on real decentralization weaving links between scientists, sociologists, economists and local communities. Experts, ecologists, biologists and sociologists should better take into account the public in their work, as well as the social context in which they are required to intervene. In the 1970s, quantitative ecology began to consider ecosystems as complex hierarchical adaptive systems, which have unique characteristics, such as resilience, multiscale communities, dissipative models, dynamics descriptors (such as non-linearity, irreversibility, self-organization, emergence, development, directivity), history, co-evolution, indeterminism or chaos (Day and Erdman 2018). As a result, anthropology and other social sciences have shown a greater interest for scientific ecology, sometimes in an enthusiastic way. It is time, indeed, to consider relationships between disciplines. A general theory is emerging, considering transdisciplinary relationships between humans and their environment (natural, social and cultural). It cannot be complete without the contribution of both ecology and anthropology to reconnect humans to their environment.

References

Baker S, Eckerberg K (2013) A policy analysis perspective on ecological restoration. Ecol Soc 18:1–17

Cannavo PF (2007) The working landscape: founding, preservation, and the politics of place. MIT Press, Cambridge, MA, pp 1–425

Crossland CJ, Bairn D, Ducrotoy JP (2005) The coastal zone: a domaine of global interactions. In: Crossland CJ, Kremer HH, Lindebaum HJ, Marshall-Crossland JI, MDA LT (eds) Coastal fluxes in the Anthropocene, vol 2. Springer, Berlin, pp 1–37

Dauvin JC, Desroy N, Janson AL, Vallet C, Duhamel S (2006) Recent changes in estuarine benthic and suprabenthic communities resulting from the development of harbour infrastructure. Mar Pollut Bull 53 (1–4):80–90

Day J, Erdman J (eds) (2018) Mississippi delta restoration: estuaries of the world. Springer, Cham

Ducrotoy JP (2010) La restauration écologique des estuaires. Lavoisier, Paris, pp 1–229

Ducrotoy JP (2017) Les milieux estuarienss et littoraux, Lavoisier. Sous-presse, Paris

Ducrotoy JP, Dauvin JC (2008) Spatio-temporal scales in estuarine conservation and restoration. Mar Pollut Bull 57:208–218

Ducrotoy JP, Kalaora B (2009) Territoires en mouvement et gouvernance: le cas de la Baie de Somme. In: Lafon X, Treyer S (eds) La Documentation Française. Collection "Agir ensemble pour le littoral". Documentation française, Paris, pp 139–150

Ducrotoy JP, Yanagi T (2008) Tools and concepts on ecological quality of coastal and estuarine environments. Mar Pollut Bull 57(1–5):1–2

Ducrotoy JP, Elliott M, Dejong V (2000) The North Sea: an evaluation. In: Shepard C (ed) Seas at the millennium. Elsevier, London

Eertman RHM, Kornman BA, Stikvoort E, Verbeek H (2002) Restoration of the Sieperda tidal marsh in the Scheldt estuary, the Netherlands. Restor Ecol 10:438–449

Elliott M, Quintino V (2007) The estuarine quality paradox, environmental homeostasis and the difficulty of detecting anthropogenic stress in naturally stressed areas. Mar Pollut Bull 54(6):640–645

Elliott M, Burdon D, Hemingway KL (2006) Estuarine, coastal and marine habitat and ecosystem restoration: confusing management and science. Estuar Coast Shelf Sci 74:349–366

Hamm L, de Croutte E, Anquetin PG, Paireau O (2004) The use of physical scale models for major maritime development schemes: the example of Port 2000. Houille Blanche-Revue Internationale de l'Eau:50–56

Holling CS (1978) Adaptive environmental assessment and management. Wiley, Chinchester

Ingold T (2011) The perception of the environment: essays in livehood, dwelling and skill. Routledge, London, pp 1–465

Jonas H (2000) Une éthique pour la nature. Desclée De Brouwer, Paris, pp 1–159

Jordan WR, Lubick GM (2011) Making nature whole: a history of ecological restoration. Island Press, London, 256 pp

Leopold A (1949) A sand county almanac and sketches here and there. Oxford University Press, New York

Livingston RJ (2006) Restoration of aquatic ecosystems. CRC Press, Boca Raton, FL, pp 1–423

Maris T, Cox T, Temmerman S, De Vleeschauwer P, Van Damme S, De Mulder T, Van den Bergh E, Meire P (2007) Tuning the tide: creating ecological conditions for tidal marsh development in a flood control area. Hydrobiologia 588:31–43

Palmer MA, Zedler JB, Falk DA (2016) Foundations of restoration ecology: outdoors & nature. Island Press, New York, 584 pp

Peterson GD, Allen CR, Holling CS (2010) Ecological resilience, biodiversity and scale. Ecosystems 1:6–18

Ratter BMW, Weig B (2012) Die Tide-Elbe – ein Kultur-, Natur- und Wirtschaftsraum aus Sicht der Bevölkerung. Hrsg. v. Helmholtz-Zentrum Geesthacht. Institut für Küstenforschung. Geesthacht, Tide Report

Sirost O (2009) La vie vu grand air. Aventures du Corps et Evasions vers la Nature. Editions PUN (Presses Universitaires de Nancy), Nancy, pp 1–270

Sirost O, Melin H, Lecoeur M, Bouillon D (2012) Projet Enfants du fleuve : Paysages vécus et paysages perçus de l'estuaire de la Seine. Projet Seine-Aval 4:223 pp

Temmerman S, Govers G, Meire P, Wartel S (2003a) Modelling long-term tidal marsh growth under changing tidal conditions and suspended sediment concentrations, Scheldt estuary, Belgium. Mar Geol 193:151–169

Temmerman S, Gover G, Wartel S, Meire P (2003b) Spatial and temporal factors controlling short-term sedimentation in a salt and freshwater tidal marsh, Scheldt estuary, Belgium, SW Netherlands. Earth Surf Process Landf 28:739–755

Temmerman S, Meire P, Bouma J, Herman PMJ, Ysebaert T, De Vriend HJ (2013) Ecosystem-based coastal defence in the face of global change. Nature 504:79–83

Vinebrooke RD, Cottingham KL, Norberg J et al (2004) Impacts of multiple stressors on biodiversity and ecosystem functioning: the role of species co-tolerance. Oikos 104(3):451–457

Ysebaert T, Fettweis M, Meire P, Sas M (2005) Benthic variability in intertidal soft-sediments in the mesohaline part of the Schelde estuary. Hydrobiologia 540:197–216

Behind Good Ecological Status, the Quest to Reconquer Water Territories

Olivier Sirost and Charly Machemehl

Summary

The end of the nineteenth century was characterised by urban man being gradually cut off from the estuarine river areas. Containment, channelling and crossing structures put water territories in the large cities at a distance. This separation was also expressed in the acceleration of the contamination of the rivers and estuaries by industry and agriculture. Leisure activities at the edge of the water were therefore relegated more to the smaller rivers or to lakes. The expansion of seaside and spa tourism turned river towns into entities distanced from the cities.

It was not until the 1950–1960s that a movement arose to reconquer and take by storm the harbour wastelands. But it was above all the European legislation on the quality of water which was to presage a reappropriation, then the programming of the urban water territories for leisure activities. The harmonisation of urban regeneration projects collided with the singular stories of river leisure activities in France. These conditions were observed in a research programme (called the Liteau Programme) funded by the Water Agencies and the Ministry of the Environment aimed at establishing good ecological status indicators for the three major Atlantic estuaries in France.

Keywords

Water framework directive · Leisure activities · Reconquer · Banks · Rivers · Estuaries · Loire · Seine · Gironde

Introduction

Rivers have a history made of human footprints and social developments (Bravard & Magny 2002). The latter should not be considered solely as the human transformation of natural spaces but also and above all as a trace of the self and evidence of society (Ginzburg 2010). The relations between people and rivers are part of a long social ecology, a way of "inhabiting an area" (Ingold 2011). In this inhabiting, we can detect the emergence of an ecological sentiment specific to human societies (Brunhes 1934), as well as a geography of the body marked, in particular, by leisure pursuits (Febvre 1922). The end of the nineteenth century in France saw the finalisation of the developments of the waterways for sanitation and industry. At the same time, we note an overcrowding of the urban centres which abandon the idea of preserving the physical integrity of the rivers and the ecological recycling of waste in particular by spreading on the fields (Barles 2005). In parallel, the health-oriented discourse describes the waterways of the major urban centres as open-air sewers that manifest the organic nature of the belly of the town to use Zola's metaphor (Serres 1975). Urban thermodynamics would seem to be more compatible with the ecology of natural spaces. Boating and fishing were growing in popularity all over Europe with the landscaping of bodies of water outside the towns, and in the streams in the countryside where sports clubs created their playing fields (Taylor 1997). A vast "back to the land" movement invited city dwellers to flee the industrialisation of the town centres and generated a veritable pastoral impulse (Marsh 1982). From the 1950s, with the delocalisation and closure of industries, a large number of wastelands were formed. At that time, in the United States, nature policies made the person practicing sports a preferential relay on the terrain within the scope of the formation of the progressive movement (Nash 1990). It is against this background that Baltimore saw its town centre and harbour areas being re-developed in the 1950s. The centre was to become an ecological neighbourhood combining heritage, parks,

O. Sirost · C. Machemehl (✉)
CETAPS EA 3832, Normandie Université, Mont Saint Aignan Cedex, France
e-mail: Olivier.sirost@univ-rouen.fr; charly.machemehl@univ-rouen.fr

C. Machemehl et al. (eds.), *Reclaiming and Rewilding River Cities for Outdoor Recreation*, Estuaries of the World,
https://doi.org/10.1007/978-3-030-48709-6_4

alternating leisure areas and "festival market places". The sea front was to convert its wastelands into tourist space, with parks, beaches and marina (Baffico 2004). In the 1970s, harbour, industrial and military wastelands afforded new opportunism in the eyes of the developers. In Britain, the transformation of the industrial wastelands along the Clyde attests to the importation of this know-how in Europe. The passion of the London gentry and the European bourgeoisie for popular culture was an invitation for the elites to take up the mechanical aesthetic, the work overall and the industrial habitat. The docks and abandoned bunkers were transformed into luxury lofts around which the urban environment was naturalised. The promenades favouring soft travel (pedestrian or bicycle), the transformation of the waterside into retail and service areas found a new harmony and aesthetic combining ecology and industrial heritage. The urban and the natural reconnected there (Barlett 2005).

From the 1970s, public water policy fell within the scope of the European Union, addressing the uses of water (drinking water, swimming, fish farming and shellfish aquaculture), then the reduction of pollution (wastewaters and agricultural nitrates). European legislation comprises around thirty directives on water.

The aim of the Water Framework Directive (WFD) is to compare territories with variable surface areas and compositions, with the ambition of applying identical criteria of ecological quality. This chapter poses the bases for reflection on the three estuaries at the crossroads of morphology and geography (spatial), culture (representations) and society (stakeholders).

The Spatial Dimension of the Estuaries

According to Louis-Alexandre Romaña, France has three major estuaries on the Manche-Atlantic seaboard.

The Gironde, whose catchment area is relatively thinly populated and not highly industrialised, has conserved a natural balance which manifest in a well-established geomorphological climax community and low pollution, which, in addition, is reinforced by its large expanse.

- The Loire has a more developed and richer drainage basin, which has led to the landscaping of the estuary, in particular, to facilitate the passage of large ships or to extract sand and gravel upstream. The morphology of the estuary has changed, causing seawater to move upstream.
- The Seine estuary is the sea outlet for the waters of a river basin which is home to 30% of the French population (including the greater Paris area) and 40% of domestic economic activity. It is faced with serious problems of pollution.

Furthermore, the importance of its role as a communication channel has been the cause of innumerable developments which have transformed it into an estuary that is completely contained. It has radically lost its "flared" shape.

The first obvious observation is that for each of the three Atlantic estuaries we are not talking of the same spatial configuration. This remark calls for a second one; the division of these territories into bodies of water also differs. And so the influence of the marine system on the river takes on different modulations depending on the expanse, the human structures and the forms of conjunction between the tide and the river flow rate. These morphological differences have heavy consequences for the perceptions and conceptions of the territories. If we take the case of the physical reality of the Seine, which is largely canalised and channelized, can we still call it an estuary? On the contrary, the vast opening proposed by the Loire makes the distinction between sea and river more perceptible (Table 1).

The aquatic footprint varies between estuaries, which is also legible in the number of villages/towns concerned by the territory. The map (Fig. 1) shows the natural and social morphologies reveal different territorial realities.

The national statistics data or the European *Corine Land Cover* programme confirm this variability between estuaries. The Seine estuary has the highest percentage of artificial surfaces, while the Loire estuary has the highest percentage of agricultural land and the Gironde has the largest percentage of seminatural areas. These overviews are confirmed in detail in Table 2.

Table 1 Water bodies, area and administrative boundaries

Indicators	Seine estuary	Loire estuary	Gironde estuary
Boundaries	Upstream, Poses Downstream, Cap d'Antifer/ Ouistreham	Ancenis at Saint Nazaire	La Réole and Génissac at the mouth
Area	50 km^2	217.6 km^2	635 km^2
Morphology	Islands and backwaters (Poses to Rouen) Pipe (Rouen to Tancarville) Mouth (Tancarville to the sea)	Double vision: Longitudinal associated with the tide Lateral combining mudflats, marshes and adjacent wetlands	Garonne and Dordogne String of islands Vast mouth (12 km wide)
Singularity	The most artificial estuary in France	Estuarine territory maintained culturally	The largest estuary in western Europe Surfers' tidal bore
Bodies of water	3	2	7

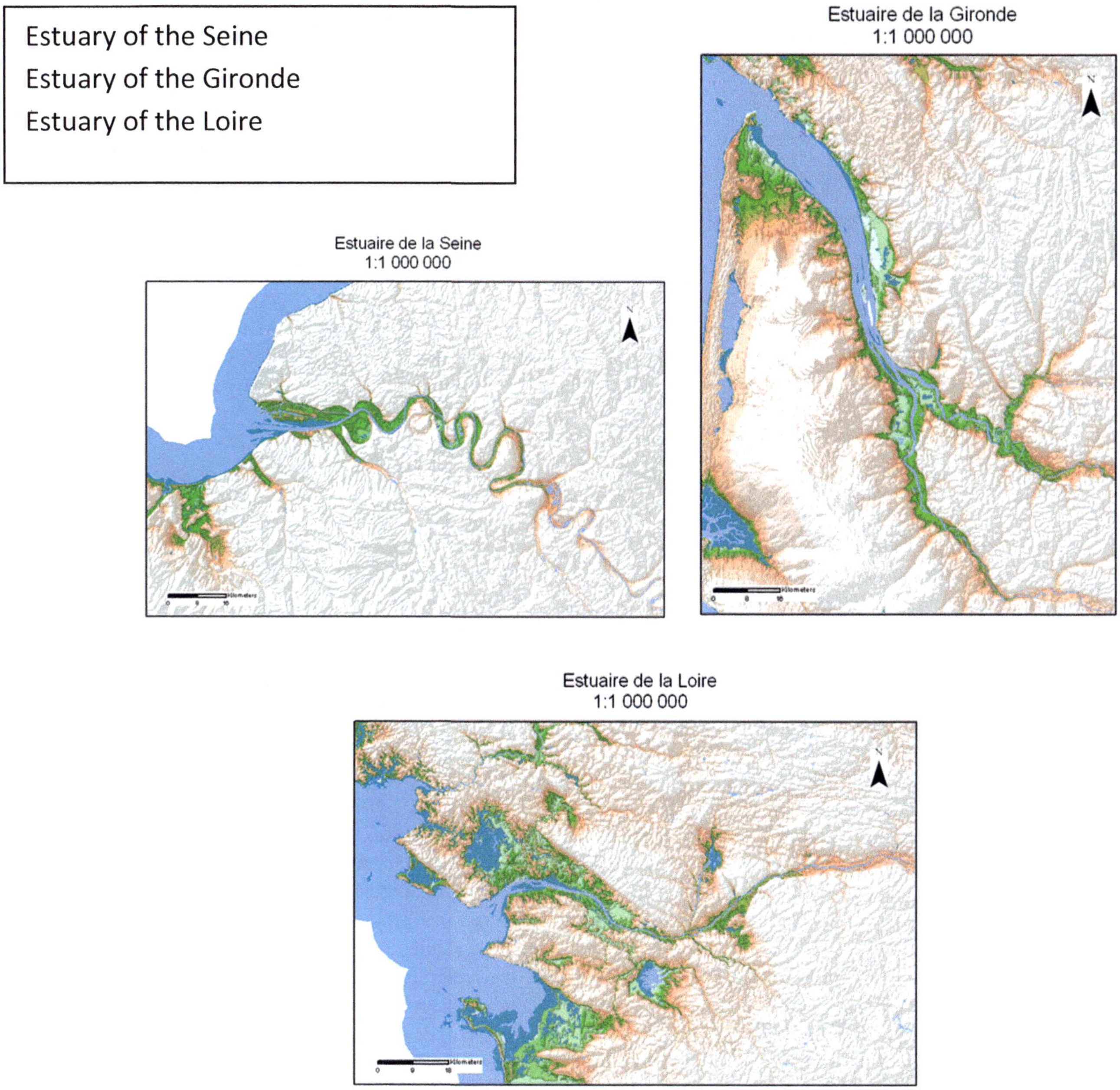

Fig. 1 Boundaries of the three major Atlantic estuaries

Once again, this spatial data clearly typifies our estuaries. The estuary of the Seine stands out as proportionally more enclosed, lacking in space because of the extent of artificial surfaces and high urban and industrial density. The Loire estuary, on the other hand, seems to be the most "natural", with its pastures and wetlands. The Gironde estuary is somewhat in between these two portraits, with much wine growing, but also beaches and sand dunes with tourist appeal clearly turned towards the sea. This geographic data corresponds to the rationales behind the social constructions of the estuaries. While forms of professional river fishing and the old occupations associated with inland waterway transport have disappeared from the Seine, in the Loire, the marshland unions, organisations of hunters, fishermen and farmers have managed to maintain a social relationship to the estuary. In the Seine estuary reed cutters, hunters and farmers remain stakeholders in the estuarine environment but their numbers are limited and their roots in the territory are less marked. For the Gironde, seaside activity, winegrowing and maize as a crop have intensified, chasing out the small trades related to fishing. Here again, traditional catch practices in the estuarine environment that are economically viable are therefore becoming rare. But this estuary also has vocations associated with the quality of life (wine, leisure activities

Table 2 Comparison of *Corine Land Cover* data for the three Atlantic estuaries

CLC data/Estuary	Seine	Loire	Gironde
Number of towns/villages	147	63	265
S^2 in water of the estuary	11.4%	13.8%	9.9%
% urban areas	11%	12%	7%
% industrial and commercial zones	6.5%	4%	1.8%
% mines, landfill sites and construction sites	0.75%	0.2%	0.32%
% green spaces	0.78%	0.82%	0.62%
% arable land	20%	11%	9.5%
% permanent crops	0.07%	1.7%	15%
% pastures	16%	26%	8%
% heterogeneous agricultural areas	5%	23%	12%
% forests	25%	2.5%	15%
% inland wetlands	0.1%	1.6%	0.85%
% marine wetlands	2.8%	2.1%	1.4%
% Natura 2000 areas	16%	25%	23%
Population density	520 people per km^2	480 people per km^2	280 people per km^2

and cuisine) unlike the estuary of the Seine that is developing along a heavily industrialised path.

The physical morphologies here undeniably reflect the economic and social morphologies. A map analysis of the land cover of the three major Atlantic estuaries illustrates the rationales that lay a framework for the description of the ecological quality (Fig. 2):

- Urban and industrial pollution for the Seine with a high level of enclosure.
- Green spaces, relatively well preserved and agricultural for the Loire, on a territory that is more flared and tightened.
- The Gironde is more beach-oriented with a vast opening at the mouth.

The Representations of the Estuaries

The Loire, the Seine and the Gironde sketch out the potential for rewilding through leisure activities of the historical paths, the reconstructions or the new usages which throw a differential light on a policy of social reconquest. Beyond the question of quality of the estuarine environment, there is a misleading assumption. The estuary entity and its territory are not a foregone conclusion for everyone. Who is talking about an estuary? The question, which may appear strange at first sight, is an excellent means of illustrating the ways the idea of ecological status is conceived of and perceived.

In the Seine estuary, the local population speaks easily of the river, with, upstream of the estuary area a certain amount of confusion with some of the tributaries (the Eure in particular). On the other hand, the population at the mouth is clearly positioned towards the sea. This absence of the estuary remains above all related to its lack of visibility and accessibility (there are few opportunities to perceive the estuary, for example, when you cross the Normandy Bridge). The estuary exists more as a line of separation between the two banks (few bridges) than as an area.

Only the players in landscaping, politics or scientific expertise use the term. The word is institutionalised in particular through the estuary Council, Grenelle (round table) or the "SOS estuaire" association which illustrates the tripartite division between scientists, developers and conservationists. However, for the scientists specialising in the Seine it has now become difficult to speak of estuary, so greatly has the area been landscaped and developed. In the period of analysis of the environment, which stretched from the Seine SAUM (*schéma d'aptitude et d'utilisation de la mer*) in 1976 to the Seine-Aval programme created in 1995, it is obvious that we are striving to conserve the ecological functions of a biotope that physically is no longer really there. This is attested by the low rate of participation of the stakeholder populations in the different public proceedings. Due to the economic interests that are given preference to the detriment of the conservation of ecological areas as illustrated by the construction of the Normandy Bridge or the Port 2000 container terminal, the citizens scarcely participate at all in the consultations where they no longer have anything to argue. On the other hand, a few estuary "beavers" including ecologists (LPO, HNNE, GONE, etc.), reed cutters, farmers, hunters, chasseurs and fishermen occupy areas of debate and influence the dynamics of compensation and rewilding of the environment.

Concerning the Loire, the word estuary is commonly used by the stakeholders and the population. It must be said that it is subject to strong socialisation operations such as estuary

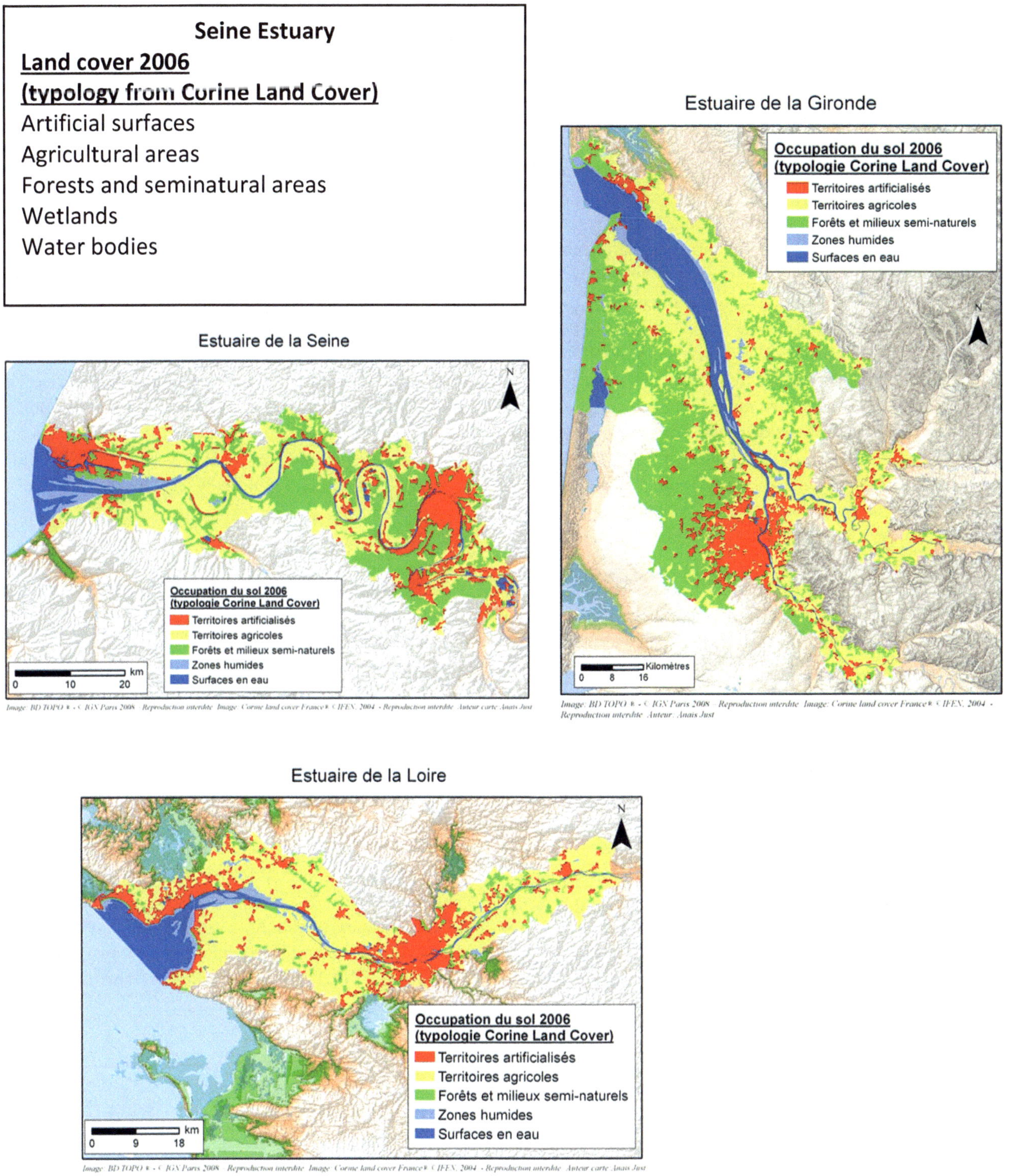

Fig. 2 Land cover 2006

classes,[1] or a contemporary art biennial. The marshland unions, which have had a strong presence since the end of the nineteenth century, the creation of the Brière Regional Natural Park in 1970, along with the historians maintaining a memory have all contributed to raising awareness of this estuarine territory. The presence, also very marked, of lively and avant-garde environmentalists from 1970 onwards and the "Loire Vivante" network has preserved the milieu from developments (for example, the Donges refinery combat). The presence of historians and geographers taking charge of the Loire PIREN conference programme and its workshop zone also explain this quite particular situation of the Loire estuary. According to the Estuarium association (formed in particular by a cultural heritage centre (Ethnopôle) in partnership with the Regional Directorate of Cultural Affairs (DRAC), the estuary became more socially structured at the end of the 1990s. The question of the word estuary and of the territory was the reason for the creation of Estuaria magazine. The estuary has become an entity around which projects crystallise: a national nature reserve planned for 2014, a communication tool for the agglomerations, "Petite Planète", île de Nantes, Saint-Nazaire Port town, river meetings, "The city of the estuary" project. In the context of decentralisation, the Pays de la Loire region is one of the three territories that accept the privatisation of water in its management, which no doubt explains the fact that it is ahead in terms of projects for the social reconquest of the edges of the water and the harbour areas.

For the Gironde, the construction of the estuary cannot be dissociated from the social reactions to industrial sites and environmental contamination. This results first of all from an institutional determination (mixed management agent for the sustainable development of the estuary: SMIDDEST) even if it was the installation of the Blayais nuclear power station that led to the launch of impact studies, contributing to the production of knowledge on the estuary, giving it cohesion. The recent mobilisations tend to reinforce this emerging identity that the segmentation of the territory had prevented from forming. The successive crises of bans on fishing (eel, shad) or shellfish aquaculture attest to this. The estuary of the Gironde has been built by the mobilisations of users against the refineries, nuclear power plant or planned liquefied natural gas terminal. These mobilisations have contributed to the birth of sanctuaries, such as the acquisitions of the Coastline Conservatory, the natural park of the estuary or the "Terre d'oiseaux" nature centre. Despite these showcases, an estuary festival or an "estuary lamb" label, the term estuary remains largely unused by the population who speak of river, sea or ocean. It is above all the water development master plan (*Schéma d'Aménagement et de Gestion des Eaux*) run by the SMIDDEST that is involved in the construction and mediatisation of the estuary of the Gironde. The silt plug is also very important for people in how they situate themselves in the estuary. Depending on the degree of familiarity, people will say they are on the ocean (downstream) because the water is clear or in the river (upstream) because it is muddy. The more informed players—experts, fishermen and interested locals—go beyond this division...

Multiple Arcs

The three estuaries can thus be contextualised on the basis of three major arcs (their social representations and services, their scientific analyses and their technical developments), which have jointly aided in their construction. It is no doubt in these divergences and resemblances that points of comparison can be found that would be useful for the WFD.

Cultural Arcs:
- Seine: Theosophical romantics, Impressionists, Pastoral writers, Ethno museums and Proliferation of associations
- Loire: Marshland unions, estuary classes, festival, Estuarium and Museums
- Gironde: Crossing on the Compostella route which was deemed very difficult in the past, Estuary Conservatory, Vitrezay nature centre, Estuary Park, etc...

Scholarly Arcs:
- Seine: Natural History Museum, Botanical Conservatories, Seine Navigation Service, SAUM, Interdisciplinary Research Programme on Water and the Environment, Seine-Aval
- Loire: Estuarium (DRAC), Loire Public Interest Grouping, Loire Estuary Measurement and Assessment Cell
- Gironde: CNRS (metallic pollution) and CEMAGREF (migratory birds)

Technical Arcs:
- Seine: Fish nurseries, Regional Nature Park, Nature Reserve, EANA Ecology Park, etc.
- Loire: Brière Regional Nature Park, Nature Reserve and hunting reserves
- Gironde: Industries at Bec d'Ambès, abandoning of the planned liquefied natural gas terminal at Verdon; marine park project, reintroduction of the sturgeon

The work conducted on each estuary shows this concomitance of arcs, in a timeline of estuarine events that are as

[1] These classes offer primary school pupils a week of visiting natural protected areas along with teaching on natural science, an exploration of the natural and cultural heritage and citizen action.

much a part of the construction of the social and natural object as of the definition of its scope and prerogatives.

The reconquering of the shorelines relies, however, on three differentiated constructions, which are part of the territorial projects, each of which has its own dynamics:

- LOIRE: "Recreate the territory" starting from the estuary.
- SEINE: Conserve certain ecological functions inherited from an estuary that no longer exists.
- GIRONDE: A conservatory of nature with multiple identities. What is most striking about the Gironde are the ideas about its preserved state very largely called into question by the collapse of the fishing stocks and the first WFD measures. Even scientists are surprised and disappointed! The estuary is much more contained, transformed and impacted than was thought!

One question: Beyond the realities of the estuaries, it is a land planning policy that is embodied in a key phrase: "users reconquer the environment", or how to recycle a space from a social perspective which resembles an inter-ministerial urban policy! The urban spread characteristic of the harbour areas and river cities has left spaces free that were once highly dense industrially. Today there are instances of opportunism to reconvert to render service to society that imply a certain privatisation of the relation to water. The WFD fully recalls this issue of the services rendered by water in the conversion of areas that were artificial surfaces and then left vacant by industry. The recreational lakes, the eco neighbourhoods, community gardens, ecology parks, beaches, the landscapes quays, marinas, green belts and blue belts... are a few of the striking illustrations of this policy of reconquering which is fully a part of a definition and monitoring of the quality of the water.

Thus, the leisure-occupation of water cities questions the corporal bond with the ecological approach as Tim Ingold suggested. And effectively the resemblances are disturbing. On areas whose histories, land cover and outdoor nature activities are very different, we are witnessing a homogenisation of the leisure development of the habitat and services (Figs. 3, 4, and 5).

Green buildings, transient beaches, sports spots on your doorstep, views of the water, areas with a patchwork of convenience retail units and cultural services, places for community gardening and DIY, and canals repurposed for pleasure boating... So many external symbols that sketch the hopes of a new way of occupying living space which the architects are starting to measure up. In this activity which is not, according to Jonas (1990), essential, it remains to be seen whether the *Homo sportivus* of the river cities will propose, not a way to restore lifestyles, but a new environmentalism drawing undreamed of resources from the wild thoughts of the body. And already to the technical leisure activities of the urban areas (Dumazedier & Ripert 1966) are reborn a whole series of celebrations of nature emphasising gardening, landscape immersion and nature watching. It is no doubt in this profound ecology that the policies for reconquering through leisure activities have a bright future before them.

Fig. 3 Nantes (Grand Elephant; Source: Waterced)

Fig. 4 Bordeaux (Water mirror quai de la Gironde, Bordeaux; Source: Pline)

Fig. 5 Rouen (Summer event "Rouen beach" on the quays of the Seine; Source: Olivier Sirost)

References

Baffico S (2004) Baltimore, une saga portuaire. Urbanités 4:1–12

Barles S (2005) L'invention des déchets urbains: France : 1790–1970. Champ Vallon, Seyssel

Barlett P (2005) Urban place: reconnecting with the natural world. MIT Press, Cambridge, MA

Bravard J-P, Magny M (2002) Les fleuves ont une histoire. Editions Errance, Paris

Brunhes J (1934) La géographie humaine. Félix Alcan, Paris

Dumazedier J, Ripert A (1966) Le loisir et la ville. Seuil, Paris

Febvre L (1922) La Terre et l'évolution humaine, Paris, Albin Michel, « L'évolution de l'Humanité »

Ginzburg C (2010) Mythes, emblèmes et traces. Morphologie et histoire. Verdier, Paris

Ingold T (2011) The perception of environment: essays on livelihood, dwelling and skill. Routledge, London

Jonas H (1990) Le principe de responsabilité: une éthique pour la civilisation technologique (J. Greisch, Trad.). les Éd. du Cerf, Paris

Marsh J (1982) Back to the land: the pastoral impulse in Victorian England 1880–1914. Faber & Faber, London

Nash RF (1990) American environmentalism: Readings in conservation history. McGraw Hill, New York

Serres M (1975) Feux et signaux de brume. Zola. Grasset, Paris

Taylor H (1997) A claim on the countryside. A history of the British outdoor movement. Keele University Press, Edinburgh

Planning and Designing Facilities that Enhance Rivers and Encourage the Development of Tourist and Recreational Spaces: Urban Promenades

Sylvie Miaux and Maxime Demers-Renaud

Summary

This text aims to present evolution of planning projects along riverbanks demonstrate the desire to enhance the river heritage as well as to develop sustainable mobility while promoting the development of tourist and recreation areas conducive to the population's well-being. So, we have decided to take an interest more specifically in the urban promenades that are being developed along rivers (Québec City, Bordeaux and Saragossa). According to what dimensions have the concepts of urban promenades developed in Québec, Bordeaux and Saragossa been built? Can we analyze these concepts in order to identify the place of nature in the city? Thus, we aim to analyze the concepts developed for each of the selected urban promenades by emphasizing the elements relating to nature and identifying the differences and similarities between the different concepts. To do so, we will first discuss urban promenades and their evolution in time. Secondly, we will describe and analyze the different promenade concepts. Finally, we will focus on the "nature" dimension of these different urban promenades.

Keywords

Urban promenade · Nature · River · Tourism · Recreational space

Introduction

While many cities in the world are reclaiming their river banks, we note increased enthusiasm for the development of sporting and leisure facilities as a means of reclaiming these spaces, which come in various forms: open sporting facilities (football fields, volleyball, basketball and skateparks), bicycle paths, urban promenades, parks, playgrounds for children, etc. This is part of a desire to enhance these watercourses while striving to meet health and sustainable development challenges. Thus, planning projects along riverbanks demonstrate the desire to enhance the river heritage as well as to develop sustainable mobility while promoting the development of tourist and recreation areas conducive to the population's well-being. In response to this new trend, we have decided to take an interest more specifically in the urban promenades that are being developed along rivers (Québec City, Bordeaux and Saragossa). According to what dimensions have the concepts of urban promenades developed in Québec, Bordeaux and Saragossa been built? Can we analyze these concepts in order to identify one of the challenges put forward by the Millennium Ecosystem Assessment (2003), namely the one related to sustainable development and more specifically, the place of nature in the city? Thus, we aim to analyze the concepts developed for each of the selected urban promenades by emphasizing the elements relating to nature and identifying the differences and similarities between the different concepts. To do so, we will first discuss urban promenades and their evolution in time. Secondly, we will describe the areas under study and the way we have analyzed the different promenade concepts. Finally, we will focus on the "nature" dimension of these different urban promenades to highlight the ways of integrating the latter.

S. Miaux (✉)
Department of Leisure, Culture and Tourism Studies, University of Quebec at Trois-Rivieres (UQTR), Trois-Rivières, Québec, Canada
e-mail: Sylvie.Miaux@uqtr.ca

M. Demers-Renaud
Environmental Planning and Design at UDeM, Université de Montréal, Montréal, Quebec, Canada
e-mail: maxime.demers-renaud@umontreal.ca

C. Machemehl et al. (eds.), *Reclaiming and Rewilding River Cities for Outdoor Recreation*, Estuaries of the World,
https://doi.org/10.1007/978-3-030-48709-6_5

Reclaiming Watercourses

Today, the enhancement and economic development of watercourses increasingly take place through tourism and leisure. Several Western cities have turned to these leisure and tourism activities (e.g. Baltimore, Bilbao and some areas of London) to combat the effects of deindustrialization. In addition, port districts have often been the most affected by industrial activities although they are often central districts (Marshall 2001). Urban and environmental revitalization carried out around the banks thus improve the quality of the spaces (Marshall 2001). The urban atmosphere associated with watercourses also naturally encourages these tourist and leisure activities.

It is therefore increasingly recognized that water is not only a useful resource but that it can also be an aesthetic and leisurely resource. Similarly, although the initial analyzes explained the interest in riverbanks and watercourses for reasons of economic change, the environmental, social and recreational benefits that can be derived from them are now being considered. The banks can thus be perceived as places favourable to tourism, but also as a symbolic "cultural capital" that will generate public spaces (Malone 1996).

Growing values associated with the environment, heritage and recreational sites are leading to a review of the value attributed to watercourses. In the context of changes in the economic structure of an industrial city to a service economy based on knowledge, leisure, recreation and aesthetics, but also in view of the growing environmental values, the revitalization of watercourses becomes a priority (Marshall 2001). Various urban promenades have been created in this context, a period when health and sustainable development issues bring new challenges to urban planners to rethink river cities and the enhancement of banks.

Urban Promenades: Evolution

This form of planning is not new; it is inspired by older forms. To illustrate the evolution of urban promenades, we will focus on the Parisian case, which has been the subject of several studies (Turcot 2007).

To begin, promenades were established in the sixteenth century as an act of civility in connection with the writings of Erasmus (Beck 2009; Turcot 2007). The destruction of ramparts allowed the development of spaces dedicated to promenades, such as the *Cours La Reine* (a planted alley inspired by the Italian model) to satisfy the nobility's ritual of social visibility according to the established rules of convention.

Contrary to this tendency, in the eighteenth century, social treaties criticized social promenades for a kind of stroll that facilitated the expression of "the truth of inner feelings" (Turcot, p. 93). In this same period, Rousseau raised awareness for the virtues of solitary strolls. Thereafter, the benefits of walking on health are put forward by Tronchin for public hygiene, among other things. This is all the more urgent at a time when epidemics imposed new ways of making the city healthier. Thus, public promenades were developed to ensure convenience, sanitation and urban aesthetics.

Boulevards appeared through this movement, marking a transition in the way people see the promenade, which became more accessible to the entire population rather than being reserved for the aristocracy. Promenades went from enclosed spaces located outside the urban forms to finally being integrated to buildings, and served as both a traffic lane and a public promenade, thus satisfying the imperatives of urban traffic (Turcot 2007). The aristocracy, who did not want to rub shoulders with the rest of the people, did not approve of this change (Beck 2009). It is for this reason that in the nineteenth century, aristocrats preferred to use carriages rather than walk. It is also in this period that *flâneurs* flooded large city streets, creating a new form of wandering that distinguished itself from the bourgeoisie, the people. In terms of planning, this tension between social classes motivated the creation of new spaces. Thus, inspired by the parks of London dedicated to the popular classes, French developers in the movement of urban Haussmanization (Beck 2009) created the Parc des Buttes-Chaumont and the Bois de Vincennes with an aesthetic, hygienic, social, moral and pedagogical purpose (covering the civility aspect: knowing how to behave appropriately). The promenades remained spaces for encounters and staging but varied (spatially) according to social class.

Then, the arrival of the automobile in the twentieth century changed the development of cities in favour of car traffic at the expense of pedestrians and strollers. The space devoted to walking was considerably reduced and interest in the promenade diminished for some time. The development of streets, highways and parking was prioritized.

Nevertheless, by the end of the twentieth century, environmental concerns, sustainable development, and health-related issues (fighting inactivity) changed the way people do, think and live the city. A true change took place in the heart of cities and their public space and took shape in different forms throughout the world: "shared space" in the Netherlands, "living street" in the United Kingdom, green corridors, planted walkway and green neighbourhoods in Paris, the "High-Line" in New York, the Promenade Samuel-de Champlain in Québec City, the "Quais" in Bordeaux, and so on. Through these numerous examples, the spaces that are set up to facilitate sustainable mobility are not limited to commuting; they are also leisure areas where users can "be at one" with the city (Lebreton and Andrieu 2012) by climbing and jumping when doing parkour or by gliding when skateboarding. This has led to the development of

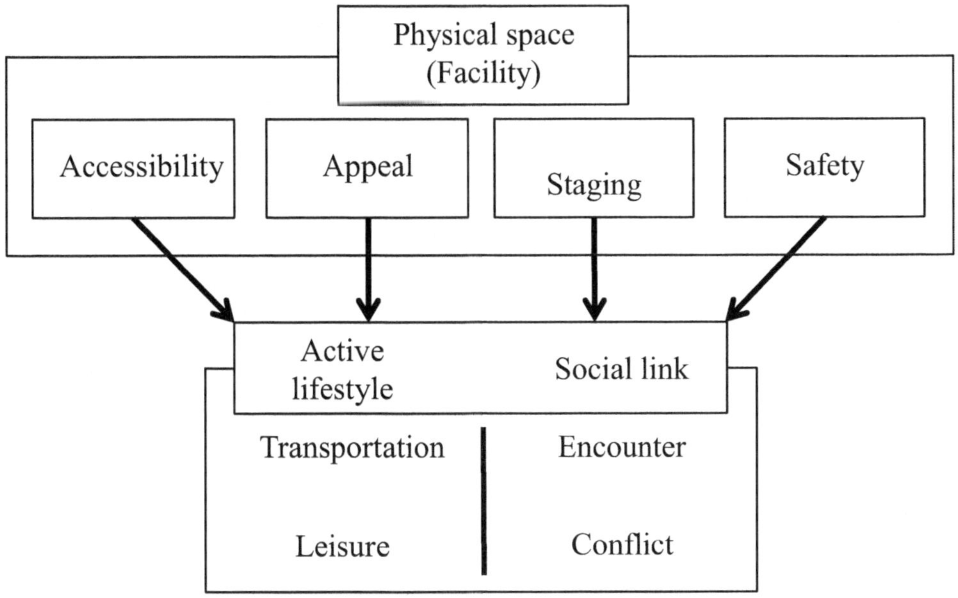

Fig. 1 Elements of the development of recreational spaces. Source: Sylvie Miaux and Maxime Demers-Renaud

new spaces such as in Copenhagen, where areas dedicated to parkour and skateboarding are arranged along the green corridor, which constitutes a promenade area that includes various sporting facilities and games accessible to all at all times.

At the turn of the twentieth century, La Rambla was the example of a promenade (a type of urban promenade located in Barcelona) and was based on three dimensions: accessibility, a very strong potential for encountering others (facilitating social ties) and the staging of otherness (Berdoulay and Morales 1999). We find the characteristics which, over time, emerged from the ways to stroll. The writings of Jan Gehl (2012) about the city on a human scale also confirmed the importance of the three dimensions mentioned above, where he insists on the city: animated (social link), sustainable, healthy (accessible to all) and safe (safety and appeal), which requires the development of slow travel (walking, cycling, etc.). Knowing that in Latin culture and more specifically in Spain, where urban promenades are anchored in history and lifestyle habits, we note that the latter is above all a place for socialization (Rieucau 2012). This is also the case in cities all over the world, where urban promenades transform inhabitants into proactive actors of a leisurely use of public space (Monnet 2012, p. 202). It is in the face of this growing demand for spaces for socialization with leisurely vocations that we will analyze the concepts of promenades from the dimensions that have just been presented (Fig. 1).

The Concept of Promenades: Analysis of Various Documents

To better understand the richness of the various forms of promenades researched, we preferred an inductive approach in order to reveal the complexity of these experiments. Thus, we have chosen to work on the analysis of various documents (11 in all) presenting the concepts of urban promenades of the cities of Québec, Bordeaux and Saragossa, which provide an overview of the way things are currently being done. The selected urban promenades are the Promenade Samuel-de Champlain in Québec City, the "Quais" de Bordeaux in Bordeaux (France) and Saragossa (Spain). All these urban promenades were built around 2008 and 2009. All three cities have a population of comparable size and are all located on the banks of a river. In fact, the Québec Urban Community has 751,990 inhabitants (Communauté métropolitaine de Québec), the urban community of Bordeaux has 720,000 inhabitants (La Communauté Urbaine de Bordeaux) and the city of Saragossa has 666,129 (City council annual report). The city of Saragossa built its urban promenade in the wake of a major event: the International Exposition of Saragossa. The City of Québec did the same for its urban promenade built in honour of the city's 400th anniversary. Finally, the construction of the urban promenade of Bordeaux was also part of a desire to reclaim the river. This choice is also

Table 1 Grid to help identify texts

Issuer	Target audience	Topic of the document
Project designer	Promoter	Planning requirement
Promoter	Project designer	Presentation of the proposed facilities
Promoter	Citizens	Promote facilities among citizens
Newspapers	Citizens	Facilities for citizens

confirmed by the fact that these promenades offer recognized and evolving tourist potential.[1]

For the content analysis, we selected a corpus of texts[2] based on the following inclusion criteria: according to the type of issuer, target audience and the topic of the document (Table 1).

To analyze this material, which brings together the point of view of the designers and planning stakeholders, we used the analysis of qualitative content in order to bring out the themes addressed in each of the documents. To do this, we chose an open and inductive approach of generalization and abstraction of the data, consisting of the coding of the principal dimensions and the selective coding of central and repetitive ideas. In this open approach, the grid is elaborated from the selected texts. This allows us to elaborate on the subject by allowing us more leeway in the analysis rather than limiting ourselves to a few reductive categories.

Nature and Urban Promenades

The design of the urban promenades of Québec City, Bordeaux and Saragossa was considered, among other things, to promote the environment and nature in different forms while integrating the leisure dimension in these spaces. The conceptual model used for the analysis consisted of four categories,[3] but in the case of enhancing nature, only three of them emerged: appeal, accessibility and staging. In addition, we will discuss how nature intervenes or does not intervene in the project (Fig. 2).

Accessibility to Nature

Designing an urban promenade, whose environmental enhancement is one of the components, requires access to the area. This type of access means connecting with the cities where they are located. This link with the outdoors is mainly discussed in the case of Bordeaux and Saragossa. In both cases, it is a matter of connecting the city to the promenade by improving public transit to avoid isolation. For Saragossa, reaching the destination is not only about having access to the space, but also how to link natural spaces with urban spaces in a form of connection between nature and culture. Secondly, from the moment users reach the promenade, they must also be able to move around inside to reach the different vegetated areas.

This second category is still specifically observable in Québec City and Saragossa, although their respective objectives are not similar. For Québec City, it is about accessing spaces that have been "renaturalized", since the promenade takes place in a former industrial zone. On the other hand, by linking natural areas, Saragossa also wishes to maintain the ecosystem connections normally present in nature. Finally, since all three urban promenades analyzed are located along a river, the latter represents a focal point, an element where both physical and visual access should be improved.

Attractiveness of Nature

Even when it is accessible, an unattractive space will not convince people to visit it. First, nature plays this role in attracting users to urban promenades. For Bordeaux and Québec City, this capacity is mainly based on a natural aesthetic developed and built by man. These two cities want to attract the population by imitating natural phenomena. In the first case, a mirror of water that fills and empties itself has been set up to resemble the movement of the tidal patterns that once occurred in the area. In Québec City, thematic piers highlight some of the natural features of the St. Lawrence River: high winds (structures reminiscent of flying birds), mist (creation of a false fog), etc. Subsequently, aside from imitating nature, the waterfronts of Québec and Bordeaux play with incentive gardens. In Québec City, the piers are large thematic gardens that highlight "the moods of the river". In Bordeaux, the presence of gardens in its facilities, which is limited to a zone of piers, reduces the monotony of the concrete space by giving it a dynamic feel created by the daily and seasonal variations of the flora. Finally, all promenades showcase works of art where nature plays a

[1] Québec, which, since the development of the Samuel-de Champlain promenade in 2008, continues to expand to soon join the city's port and the tourist area of Le petit Champlain. Saragossa, which also continues to develop the promenade on both banks of the Ebro. Finally, Bordeaux continues to develop its river banks with the construction of the Cité du vin and the development of the Chaban Delmas bridge, which allows the continuity of the promenade between the two banks of the Garonne.

[2] Articles in architecture journals, websites and book chapters.

[3] Attractiveness, accessibility, staging and safety.

Fig. 2 Promenades en images (S. Miaux)

central role. For Québec City and Saragossa, works of art connect to nature and culture, urban and nature. Art ties the two entities together. For Bordeaux, the gardens serve as a work of art.

In addition to the developed environment, the aesthetic quality of natural spaces and "free" nature is also featured to attract the population. Similarities exist between Québec City and Bordeaux, such as the presence of green spaces with vegetation (especially in Québec). The vegetation regulates temperature, making the site more comfortable. In addition to these environments, these two promenades have created gardens with a significant plant variety. For Bordeaux, these plants follow an arrangement that respects the art of gardens. For Québec City and Saragossa, urban promenades aim to rehabilitate or respect the ecosystem exchanges that prevailed when the shores were still untouched. In Québec, this represents a space that has been transformed into a marsh where an ecosystem has been re-created. For Saragossa, the approach is different because the entire promenade revolves around a relationship between a multiplicity of green spaces to conserve the ecosystems.

Staging Nature

Nature and the environment play an important role in the staging of urban promenades. On the one hand, the staging of promenades uses nature's dynamism as a central element. The gardens of the piers of Bordeaux and Québec City are active, they are evolutive. In Québec, the vegetation undulates with the wind, just as the gardens are inspired by the moods of the river. Between the gardens, the plains have a wavy shape to allude to the waves of the river. In Bordeaux, the gardens evolve according to the days and seasons through a change in colour and cycle of vegetation. These changes create a rhythm. On the other hand, the different natural spaces of the urban promenades are displayed. Each promenade has a different scenography, but the texts concerning Bordeaux explicitly express the will to stage the piers, which is less obvious in the other two promenade concepts.

Nature in Urban Promenade Projects

The project of urban promenades, as conceived by the project designers and promoters, is built around consideration of the natural environment. In this case, the information comes mainly from the promenade of Saragossa, where nature occupies a very important place. Indeed, green spaces are central to the project. However, for these stakeholders, environmental benefits are not the only indicators of a positive impact. A search for social and economic benefits must also be added, for purposes of achieving sustainable development. For Bordeaux, only the quotation "focused on plants" emphasizes the way of thinking "nature" in the project.

What Nature Are We Referring To?

After reading the content analysis, we chose to focus more specifically on the sustainable dimension (un des enjeux du Millenium 2003) of planning concepts by looking more specifically at how nature is integrated, thought and staged to provide an environment conducive to the development of varied recreational activities.

The three cases under study illustrate the current challenge for project managers and designers to build a territory where nature and city meet (Blanc 2009). Is it, as mentioned by Bonnin and Clavel, "wild nature, nature naturing; plant and animal natures, domesticated; gardened and landscaped nature, reproduced; figurative and represented nature, commemorated; urbanized nature" (2010, p. 581)? The analysis allows us to identify three ways of thinking about nature in the city:

In Bordeaux, nature is defined more in terms of aesthetics by privileging landscape quality through the development of gardens, also integrating a few trees in strategic locations without compromising the visibility of the river. The choice of planting trees is not insignificant insofar as trees are the natural elements most appreciated because of their ability to provide shade, and also how they facilitate distance perception for cyclists and walkers (Giles-Corti et al. 2005; Gobster 1995).

In Québec City, there is more of an in-between with the aesthetic dimension of nature and also a reference to symbolic nature (figurative-represented nature), the one that is staged to enhance the image of the city. This approach to designing urban recreational facilities is consistent with the idea that the disposition of elements is important (Bedimo-Rung et al. 2005), since it allows a site to be showcased (Desousa 2006).

Finally, in Saragossa, the river banks are wilder, since no development has been undertaken except for trails that are more integrated than in the case of Bordeaux and Québec. The ecosystem approach developed, especially in Saragossa is also part of the quest for well-being and "habitability" of living environments (Blanc and Lolive 2009), which involves a connection with nature. Indeed, several studies point to the positive effects of the proximity of nature on the well-being of the population (Hartig et al. 1991; Gesler 1992; etc.), which confirm the importance of considering nature in the city. Thus, when planning a promenade, one must consider the sensitive dimension of the latter by referring to environmental aesthetics (Berleant 1991, 1992), which focuses on citizens, their aesthetic experience of nature and everyday spaces (Blanc and Lolive 2009).

Planning an urban promenade in the twenty-first century requires, among other things, an emphasis on the place of nature in this device, which depends on the representations of nature of both the designers and the users. The three cases under study show the diversity of possible approaches. The urban promenade, which is not limited to a unique way of doing things, facilitates the emergence of different experiences.

References

Beck R (2009) La promenade urbaine au XIXe siècle. Annales de Bretagne et des Pays de l'Ouest. Anjou. Maine. Poitou-Charente. Touraine (116-2), pp 165–190

Bedimo-Rung AL, Mowen AJ, Cohen DA (2005) The significance of parks to physical activity and public health: a conceptual model. Am J Prev Med 28(2S2):10

Berdoulay V, Morales M (1999) Espace public et culture: stratégies barcelonaises. Géographie et cultures 29:79–96

Berleant A (1991) Art and engagement. Temple University Press, Philadelphia, PA

Berleant A (1992) The aesthetics of environment. Temple University Press, Philadelphia, PA

Blanc N (2009) Vers un urbanisme écologique ? Urbia 8:12–30

Blanc N, Lolive J (2009) Vers une esthétique environnementale : le tournant pragmatiste. Nature Science Société 17(3):268–272

Bonnin P, Clavel M (2010) Introduction: Quand la nature s'urbanise. Ethnol Fr 40(4):581–587

Desousa CA (2006) Unearthing the benefits of brownfield to green space projects- an examination of project use and quality of life impacts. Local Environ 11(5):24

Gehl J (2012) Pour des villes à échelle humaine. Les éd. Ecosociété

Gesler WM (1992) Therapeutic landscapes : medical issues in light of the new cultural geography. Soc Sci Med 34(7):735–746

Giles-Corti B, Broomhall MH, Knuiman M, Collins C, Douglas K, Ng K, Donovan RJ (2005) Increasing walking: how important is distance to attractiveness and size of public open space? Am J Prev Med 28(2S2):169–176. https://doi.org/10.1016/j.amepre.2004.10.018

Gobster PH (1995) Perception and use of a metropolitan greenway system for recreation. Landsc Urban Plan 33:401–413

Hartig T, Mang M, Evans G (1991) Restorative effects of natural environmental experiences. Environ Stud 23(1):3–26

Lebreton F, Andrieu B (2012) Quand le sport fait corps avec l'espace urbain. Loisir & Société 34(1):99–120

Malone P (ed) (1996) City, capital and water. Routledge, London

Marshall R (2001) Waterfronts in post-industrial cities. Taylor & Francis, London

Millennium Ecosystem Assessment (2003) Ecosystems and human well-being. Island Press, London

Monnet J (2012) Ville et loisirs: les usages de l'espace public. Historiens & Géographes 419:201–213

Rieucau J (2012) La promenade publique géosymbole de l'urbanité espagnole. La Rambla Nova de Tarragone EchoGéo 22. https://doi.org/10.4000/echogeo.13252

Turcot L (2007) Le promeneur à Paris au XVIIIe siècle. Le Promeneur

Outdoor Leisure Activities at Odds with the City? Arcachon Bay and the Massif Des Calanques

Ludovic Ginelli

Summary

Outdoor leisure activities in two "Meccas of nature"—Arcachon Bay and the Massif des Calanques—located in the outskirts of cities are currently undergoing similar processes. Due to the secular distinction between town and country, the opposition with social compulsive time is amplified in the case of outdoor leisure activities by a spatial opposition. For these users, a "double border" is traced between the urban space-time of the daily routine and the liberating activities performed in "wilded" environments. This constitutes a powerful framework for social–environmental experiences associated with outdoor leisure activities, but it is not immutable. Users are frequently disturbed by the difference between how they imagine wild nature and the experience of an environment porous to the influences of the city and its social challenges. Could this disturbance be a sign of less divided urban–rural relations, or even a reinvention of the town by outdoor leisure? The answer must be nuanced.

Keywords

Outdoor leisure activities · Experience of the environment · Ecologization · Arcachon Bay · Calanques

Introduction

Since the 1960s, outdoor activities have been increasingly practiced in many so-called developed countries. Beyond the relaxation aspect, these usages, which are now described as "outdoor sports and leisure activities", are sought out for the special experience of the environment they procure (Pröbstl et al. 2009). Based on surveys conducted around Arcachon Bay and the Massif des Calanques,[1] we shall analyse them as social–environmental experiences[2] which are part of an opposition/compensation dialectic with the compulsive space-times of users, the majority urban-dwellers, living in proximity to Bordeaux (for users of Arcachon Bay) or Marseille (for users of the Calanques) (Ginelli 2017). The activities studied here—bow hunting, sea kayaking, ancient hunts, underwater hunting—differ in their historical constructions, features of players, techniques and whether or not natural resources are extracted. For all that, the convergences seem today to dominate. In this respect, the recent requalification of these uses as "outdoor sports and leisure activities" is particularly revealing of significant transformations granted in the sense of these practices in a natural setting (Bessy and Mouton 2004) and on our study sites, it combines with the social–historical construction of the Calanques and of Arcachon Bay into "Meccas of nature". Under the effect of these two processes, such outdoor recreation is now mainly practised by urban dwellers on the edges of the city, both for the preferential experience of "the great outdoors", the know-how and the social skills they arouse among the "passionate". Thanks to this trend, these practices and experiences now directly question the town, or even all of the compulsive space-times (professional and domestic spheres), with regard to which they define themselves in opposition. To what extent can these leisure activities nourish and renew "the experience of public problems" (Céfaï and Terzi 2012) of the town and the environment?

We shall return, first of all, to the secular construction of the urban–rural opposition in the Massif des Calanques and Arcachon Bay (part 1). The outdoor leisure activities

L. Ginelli (✉)
ETBX, INRAE Cestas Cedex, France
e-mail: ludovic.ginelli@inrae.fr

[1] 73 interviews and 43 observations were conducted with users, clubs and managers of protected areas, to which can be added the participation at consultation meetings, analysis of official documents or products by users collectives (blogs, open letters, petitions, etc.).

[2] This term is taken to mean the transactions that unite the individual and social groups with a physical, social and historical environment (Céfaï and Terzi 2012: 11).

C. Machemehl et al. (eds.), *Reclaiming and Rewilding River Cities for Outdoor Recreation*, Estuaries of the World,
https://doi.org/10.1007/978-3-030-48709-6_6

examined are a part of these dialectic relations, and are today largely built adjoining the town while at the same time being practiced mainly by urbanites on the edges of the city (part 2). Given this trend, the cognitive and normative frameworks of these social–environmental experiences are frequently "disturbed" and can lead to the criticism of lived social–environmental challenges, with tension between confinement and publicising (part 3).

Arcachon Bay and Massif des Calanques, Wilded Spaces at Odds With the Town

Up until the mid-nineteenth century or even beyond, our two study sites were areas on the margins of the cities, almost deserted or occupied by a population itself marginalised. Their requalification as Meccas of nature is a result of the general process of rewilding[3] of rural areas.

Marseille and the Calanques: Once a "Desert", Now a National Park on the Outskirts of the City

The site of the Calanques, which became a national park in 2012 and attracts approximately two million visitors a year, stretches to the south of the agglomeration of Marseille, which has a population of 1.5 million. This visitor appeal is in stark contrast with the historical vocation of these semi-arid limestone cliffs of the site, chosen in the nineteenth century to locate dangerous industries (quarries, chemicals and metallurgy) away from the town (Melin 2012: 158). Until the wake of World War II, the Calanques were even used in places to relegate populations passing through (Hérat 2012). There is evidence of the "hill" and cabins being frequented by the popular categories as domestic extensions (Masséna-Gourc 1994) but it is far from being valorised like today. Why?

In parallel, starting from the 1880s, excursionist clubs from Marseille built and defended the massif des Calanques as a framework for regenerating outdoor activities and against the extension of the town and its miasma. They contributed widely to re-wilding the place and defining its legitimate uses (Masséna-Gourc 1994; Ginelli et al. 2014). As early as 1910, their militant actions prevented the opening of new quarries in the Calanques then the recurring projects of "Calanques road". These associations joined together in 1965 in a Committee for the defence of natural sites (Cosina) and succeeded in having the site listed in 1975. In 1991–1992, a vast mobilisation of users stopped a municipal project to partially develop the site (constructions on the coastline, extension of a sports centre) and reactivate the project for a Calanques national park. Finally created in 2012 after a procedure that was consultative and contested (Deldrève and Deboudt 2012), the park further accentuated the wilding of the Calanques and extended their maritime section. The zoning, combined with the recent increase in density around Marseille (Claeys et al. 2012) and the city's policy regarding the surroundings of the Calanques, reinforced the town and country division with negative effects in terms of opening to the public and equal access. Intertwined territories—in particular the Calanque de Sormiou and the underprivileged neighbourhood of La Cayolle—were separated, hampering practices (bathing and rambling) that allowed people to own the space and escape temporarily from a difficult urban situation (Deldrève and Hérat 2012). This process also partially explains the high mobilisation of more influential users for the creation of the national park such as ramblers, cottage owners, climbers and hunters (Deldrève and Deboudt 2012).

Arcachon Bay: From "Miserable Country" to Marine Natural Park on the Outskirts of the City

Our second study site, Arcachon Bay, is a lagoon situated some 60 kilometres from Bordeaux. It is edged by an attractive coastal conurbation with a population of more than 100,000,[4] highly touristic[5] and directly connected to the Bordeaux agglomeration (almost a million inhabitants) by rail and road. In parallel to this tourism and residential appeal, the professions related to the exploitation of natural resources (fishing and oyster farming) are regressing.[6] As in the Calanques, the trajectory and dynamics of this territory result from a secular process of wilding.

At the beginning of the eighteenth century, Arcachon Bay was perceived as a largely isolated and unproductive space or even a "miserable country" (Auly and Veiga 2010: 151) which remained to be developed. A large number of landscaping, sanitation and planting projects were conducted: major containment works to create salt marshes and pine plantations along stretches of moorland and coastal dunes. In the nineteenth century, the exploitation of natural resources in Arcachon Bay was stepped up (Weulersse

[3] For a critical reading of the re-wilding of areas occupied by dominated populations see (Chamboredon (1982, 1985) or Sirost (2009: 35–37) in France and Cronon (1996) in the United States.

[4] After an increase of more than 25% between 1990 and 2007, the largest in the département.

[5] Arcachon Bay represents 27% of the number of stays in Gironde and a third of the tourism in the department (or almost 8.5 million overnight stays). Source: Aquitaine regional tourism committee, 2011.

[6] Source: Le Berre, S., Courtel J., Brigand L., (2010) Etude de la fréquentation nautique du bassin d'Arcachon, Géomer/ Direction régionale des affaires maritimes d'Aquitaine, 97 p.

1928). Commissioned by Napoléon III, Victor Coste invented oyster farming in 1852 and the salt marshes were converted into fish ponds and pastureland.[7]

In parallel, holiday resort tourism began in Arcachon thanks to the arrival of the railway in 1841. Spa treatments, initially praised for their medicinal virtues, were the crucible for the first aesthetic descriptions of places (Bercovitz and Briffaud 2013). Arcachon became one of the first resorts where the bourgeoisies of Europe invented society tourism and reclassified places as "wild" and "exotic".[8] After World War II, fish farming and livestock breeding declined and disappeared in the 1960s–1970s.[9] Since 1980–1990, oyster farming has also greatly declined.[10] Concomitantly, the construction of Arcachon Bay as a protected natural space was engaged, not without conflicts, and concerned first of all the "Meccas" such as île aux Oiseaux (André-Lamat and Mellac 2014), a site listed since 1943 and the Banc d'Arguin, a nature reserve since 1972. From the 1980s, numerous contained marshes were acquired by the French coastal protection agency and sometimes transformed into protected areas (salt marshes of Arès and Lège-Cap Ferret nature reserve, Domaine of Certes). The management plans of these spaces aim at "rewilding" the spaces (restoring ecological continuities, refilling the previously contained marshes with water, eliminating plant species deemed invasive—Photo 1) and removing signs of the "consumerist bulimia of the 1960s" (manager of a nature reserve, Arcachon Bay).

Created in 2014, the marine natural park extends the rewilding to the whole of Arcachon Bay, including the sea part, and aims to coordinate actions on an area that has become a "millefeuille of environmental zoning" (André-Lamat and Mellac 2014).

The current secular rewilding of Arcachon Bay and of the Massif des Calanques, carried along by the rallying of non-profit groups and instituted by nature politicians influences both the uses of the town and of nature. As in the country around Brest (Le Guirriec 2012) or in the Seine estuary near Rouen (Féménias et al. 2011), a functional specialisation is emerging along with dialectic relations and tensions/complementarities between the urban areas (agglomerations of Bordeaux and Marseille) largely associated with the compulsive space-times and formerly rural, productive areas that have become attractive outskirts, invested in for the environmental experience they procure, an antidote for urban ills.

Outdoor Leisure Activities and Daily Social Worlds, a "Double Border"

Between "Tradition" and Urban Investment in Nature: The Hunting of Migratory Birds

In our societies, even the ancestral uses of nature (hunting, fishing and gathering) are redefined according to the urban rationales, lifestyles and expectations of their adepts. "*If more and more citizens are involved, it is indeed to try to escape from the developed space-time of the cities*" (Larrère and De La Soudière 1985: 217) where contact with the natural elements is no longer daily. Certainly, the case of the oyster farmers–hunters identified in Arcachon Bay is an exception to this dominant logic. Sometimes called "sea peasants", these oyster farmers can still prolong contact with a natural milieu, by chasing waterfowl, which it is a question of mastering as in their profession (Ginelli 2015a). But for the majority of the hunters in these areas, in the majority city dwellers, hunting in huts (Photo 2) is no longer the extension of a rural, daily job. It meets a "need for nature" satisfied at times (holidays, weekends, retirement) and in places that are clearly distinct from those of work, the daily routine and urban lifestyles:

> Just the fact that you are in the country. (...) We need this. There are times, I can't leave here, when I come to make my "bricks"[11] at the height of the summer... (retired supervisor, 55, lives on the outskirts of Bordeaux)

Of popular origin, the hunting of migratory birds is mainly practiced today by men from modest social class (workers and employees) living in industrial regions (Mischi 2008; Raison du Cleuziou 2009). Within these communities of "passionate" people, it is a question of proving one's control of the environment by rallying the knowledge placed at the service of the game of the hunt:

> You have to be in osmosis with the milieu, with the bird, that is to say you have to anticipate the places the birds might alight. Therefore you are systematically in the process of modifying your technique. Then when it works, you feel like a king! (38 years old, landscape gardener from near Marseille).

[7] These agricultural and fish farming productions, to which can be added oysters and wild ducks, sources of revenue for the inhabitants, were sold for the main part on the markets of Bordeaux (Auly and Veiga 2010: 175).

[8] Source: Dupuyoo J. (2005). Autrefois... Le bassin d'Arcachon. Pêcheurs, ostréiculteurs et chasseurs. Publisher: Confluences.

[9] Sources: Plan de gestion de la Réserve naturelle des prés salés d'Arès et de Lège, et du site des « Abberts », 2010–2014, Tome 1, p. 68. and Morel P, Labourg, P.J (1988) « Zones humides du bassin d'Arcachon: synthèse des connaissances, activités, impacts et planification spatiale », Institut universitaire de biologie marine d'Arcachon / DRAE d'Aquitaine, 146 p.

[10] The number of oyster farmers was divided by five from 1960 to 2009 and the surface area of the oyster farms shrank by almost 57% over the same period (Auly and Veiga 2010: 201).

[11] "Mattes" are clay bricks made with a shovel on areas generally close to hunting facilities.

Photo 1 "Eco-citizen" worksite managing baccharis, an "invasive" shrub in Arcachon Bay (L. Ginelli)

Through the practice of these ancient forms of hunting willingly described as "traditional", these users become a part of leisure, idealised spaces of fair competition between equals, at a remove from the inegalitarian mercantile relations that characterise their daily lives.

Bow Hunting, Underwater Hunting and Sea Kayaking: Social–Environmental Experiences Against an Urban Daily Routine

Invented in Marseille in the 1930s, underwater hunting fosters masculine social skills centred on use, willingly described as "family" or "fraternal" in which, according to an egalitarian ideal characteristic of "sportivised" usages, social hierarchies are supposedly abolished. These experiences among the initiated, lived on a mode of distancing and intensification of daily life (Griffet 1999) are the subject of emotional and identity-related investments of first order. They then become resolutely irreconcilable with certain protective measures (here the non-fishing areas in the Calanques national park) deemed exclusive to humans:

> It is a sports passion. For me it is a way of life. Because on this earth, and here particularly in the cities, what is it? The advantage of having better than the others! That has never interested me. I prefer to be alone on the water with my fish! (...) Then there are people who draw a line on a map and take my garden away. I almost died for my garden! (53, employee, Marseille).

In Arcachon Bay, the bow hunters, often native to the area and employed in the secondary or tertiary sector, do not materially experience the environment every day. The attraction of hunting allows them to believe—temporarily—in the immersion in a wild nature close to the American myth of the *wilderness*. For one of the interviewed, sea kayak and bow hunting respond to the same desire to be included in wild nature by renewing with immemorial practices. This is far from an isolated case. This suggests that in spite of very different historical constructions and social origins, practices that "the advent of leisure activities" (Corbin 1995) had already brought closer together are converging increasingly by their social experiences of the environment. Given the homogenisation of their audiences, their democratisation must nonetheless be put in perspective. They would rather reveal the distribution of relations to nature recognised as legitimate—"proper uses"—than a relationship to nature shared democratically (Ginelli et al. 2014).

The users considered here invest the Calanques and Arcachon Bay as Meccas of nature, opposite from every point of view to the "compulsive space times" (Le Guirriec 2012) and the "social burden" of their daily lives (Lebreton et al. 2008). From their point of view, a "double border" emerges between their place of residence and their daily lifestyle and the liberating activities accomplished in the "great outdoors" of the Calanques and Arcachon Bay. This socially constructed division is anchored all the more in their experiences as the re-wilding of these two spaces is more than

Photo 2 Hunting huts on outside Arcachon (L. Ginelli)

a hundred-years old and accentuated by the recent creation of protected spaces.

From the Need for Nature to Social–Environmental Criticism?

Cognitive Passions of the Environment

Outdoor leisure activities, which are the target of considerable cognitive, sensitive and identity-related investments, point to major tensions that traverse our societies: urban concentration and the related lifestyles, the acceleration of social time and the influence of management systems on societies and the environment combine and render such social–environmental experiences increasingly rare. Around them, people built social skills and regimes of specific know-how, or "cognitive passions" (Roux et al. 2009). They mobilise knowledge of the places, the environment and animal behaviour in the perspective of a game with the living to feel the sensation of being immerged in the *wilderness*, sharing it and being acknowledged among fellow nature lovers. This quest for wildness, a register inherited from the first American recreational associations then the French ones in the nineteenth century and relatively democratised today, "renews and displaces the human gaze towards horizons that differ from daily reality" (Krieger et al. 2017: 12). Also, these cognitive passions are often an anchorage point for wider attention to the local environment and its changes, in other words "awarenesses of the milieu" (Féménias et al. 2011). They can open the way to criticism or even the publicising of problems that the users experience. When certain observations seem abnormal, in other words "disturb" the familiar milestones of their social–environmental experiences, the users draw up conjectures that they discuss in their networks (friends, specialist Internet forums). From the pragmatist point of view we retain (Céfaï and Terzi 2012; Ginelli 2015b), this "disturbance" may announce survey activities, which themselves are favourable to the training of a collective around new "affordances", that is to say a collective agreement on a certain interpretation of reality. This sequence of disturbance/survey/affordance queries the opposition instituted between town and country and can lead to a revision of judgements, environmental practices or even methods of organising the town and rural policies. For all that, this sequence is far from systematic, as the rationales of the survey can be hampered: as we shall see, and not surprisingly, not all experiences lead to the construction of new public problems.

Numerous users are "disturbed" by unusual observations in relation to their passion. For example, the hunters say the changes in the migration of birds flying over Arcachon Bay and the Calanques (lateness, changes in migration routes) are due to climate change and the urbanisation of their hunting territories. This primarily hunting interest in the migration of the birds is frequently a point of anchorage for a wider attention to the environment, for example about the impact of pesticides on the fauna as a whole:

> In the fields, everything is polluted! You see the seeds red with pesticides, the finches are dying in their thousands and nobody is talking about it. The only people who talk about it are those it affects.(bird hunter, 74, retired policeman, Marseille).

Observing Arcachon Bay, hunters of migratory birds and kayakers very quickly noted the depletion of the eelgrass beds, which play an important role for this marine ecosystem. Some have become "whistleblowers" about the phenomenon, at the origin of two research programmes[12] and protective measures. Generally, the observation of the environment, technically indispensable to their practice, arouses a lot of concerns. The proliferation of "rocks" of wild oysters deemed invasive, dangerous (risk of cuts) and obstructing the channels, the accelerated silting of Arcachon Bay, the considerable increase in nautical sports all harm the imagination of a great outdoors preserved from the town. Thus, a kayaker from Arcachon feels the shift between his ideal of "the great outdoors" and the experience of an aquatic milieu threatened by the incursions of the town and therefore finally subject to the same issues:

> There were jet-skis, fishermen, boats, yachts, a whole world of usual pleasure boaters and there were snorkelers as well. It was incredible the way the space was busy! I think that at less than 30 metres from the surface, the bay was full up... And hey... we are more or less aware of being part of it, so we say to ourselves why us and not them? (kayaker and naturalist, 50, teacher at Andernos).

Whatever the activity, many note and discuss regularly among peers the changes that affect their environments of practice categorised as "nature" and relate them to largely publicised environmental problems (pollution, climate change and consequences of urbanisation). But this environmental disturbance and ecologised common sense do not lead systematically—far from it—to a collective ecological "survey" in the sense understood by the pragmatist authors. How can this be explained?

Environmental Exemplarity, a Shared Norm... at the Risk of Overkill

Some users draw all the normative consequences from the contemporary ecology imperative by re-questioning through this lens their practices and those of others. In relation to their peers, they have the particularity of being part of the networks where the environment is central and apprehended as "ecology" in an expert manner ("environmentally responsible" naturalist associations, professional ecologists, networks of hunting experts, hunters on land and underwater).

On the other hand, the other users, part of other social milieus, confirm the ambivalence of the adepts of outdoor leisure towards ecologisation, observed on other case studies (Krieger et al. 2017). Those interviewed show themselves to be split between the social imperative of environmental exemplarity and rejecting its normative aspects, individualising and expert, in particular, because the associated management mechanisms could disqualify their environmental knowledge and hamper the social–environmental experiences they seek. The pressure to be environmentally exemplary also provokes its own splits, including within user collectives. It is particularly true in the protected areas (Arcachon Bay nature reserve, Calanques national park), where practical experiences and the sensitivities of the sports enthusiasts are necessarily placed in parentheses. It is only by conforming to the registers of expertise that the non-specialists can make themselves heard, and therefore without encouraging real changes in the collective of the participants, nor in the manners of formulating the challenges and debating the responses to be provided.[13] Even the users close to the actors in the expert management of nature by their social profile and networks (kayakers–naturalists and underwater hunter biologists, bow hunters) struggle with the injunction to be individually responsible for the environment and contest it—expertly—when it offends their sense of environmental justice. For example, when the national park was created, an underwater hunters' association committed on a national level towards "eco-responsible" goals posed a sine qua non condition: to be environmentally effective and socially equitable, they explain, the fishing restrictions must apply to all types of fishing, even to all uses of the sea.

Conclusion

With different historical backgrounds, audiences and techniques, the outdoor leisure activities examined here are today undergoing similar processes in two "Meccas of nature" on the outskirts of cities, Arcachon Bay and the Massif des Calanques. The secular division between town and country becomes a "double border" between urban compulsive space-times and leisure activities for evasion in "rewilded" surroundings. It constitutes a powerful framework for social–environmental experiences associated with

[12] These queries expressed as well by the population as by the scientists are one of the justifications of the research project of "Oysterfarming and Quality of Arcachon Bay" funded by the Aquitaine Region, to which we contributed (Le Floch and Ginelli 2013).

[13] Similar processes were observed in relation to another outdoor sport; surfing. The professionalisation of the leaders of the *Surfrider foundation Europe* and their growing insertion in public action networks makes more use of specialist skills and know-how, to the detriment of the testimony register. The association succeeds in gaining recognition as a legitimate public contact, at the cost of an intenral split between specialists and surfers who hold "situated" knowledge (Weisbein 2015).

outdoor leisure activities, but is not immutable. The users are frequently "disturbed" by the difference between how they imagine wild nature and the experience of an environment porous to the influences of the town and its social challenges, of the environment (pollution and climate change) but also of society (access for the public to natural spaces, governance, environmental justice, etc.). Could this disturbed state be a sign of less divided urban–rural relations, or even a reinvention of the town by outdoor leisure? The answer must be nuanced.

In inner Marseille, management of the urban parks is currently endeavouring to re-wild the nature in the town. But the attempts to go beyond the town–country division with green belts and blue belts or the recent recognition of the "peri-urban" nature of the national park, come up against "the imperatives of an urbanisation erected as an engine driving territorial revitalisation" (Claeys et al. 2012: 80) and a powerful densification of the outskirts of the city. Despite their recent participatory turn, nature policies remain in the grips of the naturalist tradition, to which can be added the increasing weight of expertise. Against the apparent environment-friendly consensus, we observe that greening initiatives, particularly in protected spaces (Calanques national park, Arcachon Bay nature reserve) come up against some of the users, frustrated in their sports passions; they, therefore, fragment the users' groups and their relations with the environment more than they unite them. By fully mobilising, the principle of responsibility and the register of expertise, these normalisations focus attention—and action—on uses and ecological evaluation of their impact,[14] but without really questioning the lifestyles of the different social categories (Comby and Grossetête 2012), nor the path of development of societies. When individuals resist environmental norms perceived as external to their social world, are they opposing only a rule or a particular norm, or are they also defending an autonomous space-time reversed in relation to their daily routine? Such stances, that promote the experience and control of a social–natural environment, deserve in our opinion to be interpreted—and taken into account—as instances of distancing of the omnipresence of the expert and management systems characteristic of our societies, our cities and our relationships to the environment.

References

André-Lamat V, Mellac M (2014) L'île aux Oiseaux, appropriation conflictuelle d'un haut lieu du bassin d'Arcachon. L'Information géographique 78(2):62–84

Auly T, Veiga J (2010) Le Bassin d'Arcachon, un milieu naturel menacé ? Confluences/Fédération départementale des chasseurs de la Gironde

Bercovitz R, Briffaud S (2013) Histoire des représentations sociales de la qualité des paysges et du milieu (XVIII–XXI siècle). Contribution au projet "Ostréiculture et qualité du Bassin d'Arcachon". Rapport, 27 p., CEPAGE

Bessy O, Mouton M (2004) Du plein air au sport de nature. Nouvelles pratiques, nouveaux enjeux. Les cahiers Espaces 81:13–34

Céfaï D, Terzi C (2012) L'expérience des problèmes publics. Editions de l'EHESS, Collection Raisons pratiques

Chamboredon J-C (1982) La diffusion de la chasse et la transformation des usages sociaux de l'espace rural. Etudes Rurales La chasse et la cueillette aujourd'hui 87–88:233–260

Chamboredon J-C (1985) La naturalisation de la campagne: une autre manière de cultiver les simples ? Dans Protection de la nature: histoire et idéologie. De la nature à l'environnement, ed. Cadoret, L'Harmattan, pp 138–151

Claeys C, Consales J-N, Barthélémy C (2012) Marseille et ses natures: perméabilites spatiales, segmentations sociales. Desenvolvimento e Meio Ambiente 26:69–85

Comby J-B, Grossetête M (2012) "Se montrer prévoyant": une norme sociale diversement appropriée. Sociologie 3:251–266

Corbin A (1995) L'avènement des loisirs, 1850–1960. Flammarion, Paris

Cronon W (1996) The trouble with wilderness: or, getting back to the wrong nature. Environ Hist 1:7–28

Deldrève V, Deboudt P (2012) Le Parc national des calanques. Construction territoriale, concertation et usages. Quae, Versailles

Deldrève V, Hérat A. (2012) Des inégalités garantes de la protection des Calanques ? Vertigo – la revue électronique en sciences de l'environnement 12, 2 [en ligne], Consulté le 16 août 2017. http://vertigo.revues.org/12700; https://doi.org/10.4000/vertigo.12700

Féménias D, Sirost O, Evrard B (2011) Les loisirs nautiques dans l'estuaire de la Seine. Médiations territoriales, consciences du milieu. VertigO-la revue électronique en sciences de l'environnement, Hors-série 10. [en ligne]. Consulté le 16 août 2017. http://vertigo.revues.org/11576; https://doi.org/10.4000/vertigo.11576

Ginelli L (2015a) Loisirs de nature et environnement. Spécificités des épreuves, liens avec les professions. In: Arpin I, Bouleau G, Candau J, Audrey R (eds) Dans Les activités professionnelles à l'épreuve de l'environnement. Octarès, Toulouse, pp 55–72

Ginelli L (2015b) Nature leisure activities put to the environmental test. A pragmatic, sociological approach. In: Dissart J-C, Dehez J, Marsat J-B (eds) Dans tourism, recreation and regional development: perspectives from France and abroad. Ashgate, Paris, pp 107–121

Ginelli L (2017) Jeux de nature, natures en jeu. Des loisirs aux prises avec l'écologisation des sociétés. Collection Ecopolis. Peter Lang, Paris

Ginelli L, Marquet V, Deldrève V (2014) Bien pratiquer la nature... pour protéger les Calanques ? Ethnol Fr 44(3):525–536

Griffet J (1999) La mise en commun de la nature. Du "bon coin" en chasse sous-marine et en escalade. Dossier de la revue de Géographie Alpine 20:41–46

Hérat A (2012) Une politique de nature à l'épreuve du territoire: le quartir de la Cayolle et la calanque de Sormiou (Marseille). In: Deldrève V, Deboudt P (eds) Dans Le Parc national des calanques. Construction territoriale, concertation et usages. Quae, Versailles, pp 53–72

Krieger S-J, Deldrève V, Lewis N (2017) Écologisation des loisirs de nature, entre ensauvagement et domestication. Loisir et Société/Society and Leisure 40(1):25–38

Larrère R, De La Soudière M (1985) Cueillir la montagne. Plantes, fleurs, champignons en Gévaudan, Auvergne et Limousin. La Manufacture, Lyon

[14] Taking the case of the Camargue huts destroyed disputably in the name of the environment, Picon (2008 [1978]: 214–222) points out the major risk of obscuring by such—symbolic—measures, frontline environmental threats, which are patent locally (industrial risks, flooding).

Le Floch S, Ginelli L (2013) La "qualité du milieu" à l'épreuve du quotidien. (Re)constructions ordinaires de l'idée de qualité relative au bassin d'Arcachon. Résultats d'une enquête par entretiens auprès d'usagers et d'acteurs locaux. Rapport, Irstea, 46 p

Le Guirriec P (2012) De l'espace à loisir à la pointe de la Bretagne. Ethnol Fr 42(4):689–696

Lebreton F, Héas S, Bodin D, Robène L, Abdelmalek AA (2008) Terre et ciel: étude sociologique d'espaces-temps sportifs marginaux. Espaces et Sociétés 132–133(1–2):209–222

Masséna-Gourc G (1994) Sur nos besoins de nature: l'exemple du Massif des Calanques. Forêt méditerranéenne 15(3):289–306

Melin H (2012) De l'exploitation des ressources à la protection: une histoire humaine de la nature dans les calanques. In: Deldrève V, Deboudt P (eds) Dans Le Parc national des calanques. Construction territoriale, concertation et usages. Quae, Versailles, pp 153–172

Mischi J (2008) Les militants ouvriers de la chasse. Éléments sur le rapport à la politique des classes populaires. Politix 83(3):105–131

Picon B (2008 [1978]) L'espace et le temps en Camargue. Actes Sud, Arles

Pröbstl U, Elands BHM, Veronika W (2009) Forest recreation and nature tourism in Europe: context, history, and current situation. In: Bell S, Simpson M, Tyrväinen L, Sievänen T, Ulrike P (eds) European forest recreation and tourism: a handbook. Routledge, New York

Raison du Cleuziou Y (2009) L'invention de la "chasse durable". Genèse d'une appropriation subversive des normes de développement durable par les chasseurs de la baie de Somme. In: Villalba B (ed) Dans Appropriations du développement durable: émergences, diffusions, traductions. Septentrion, Villeneuve, pp 341–362

Roux J, Charvolin F, Dumain A (2009) Les "passions cognitives" ou la dimension rebelle du connaître en régime de passion. Premiers résultats d'un programme en cours. Revue d'anthropologie des connaissances 3(3):369–385

Sirost O (2009) La vie au grand air ou l'invention occidentale des milieux récréatifs. In: Sirost O (ed) Dans La vie au grand air. Aventures du corps et évasions vers la nature. Presses universitaires de Nancy, Nancy, pp 7–44

Weisbein J (2015) Capter et (co)produire des savoirs sous contraintes: le tournant expert de Surfrider Foundation Europe. Politix 111 (3):93–117

Weulersse J (1928) Le bassin d'Arcachon. Annales de Géographie 37:407–427

"On the Conquest of Wild Nature"... But What Is Meant by "Nature"?

Sarah-Jane Krieger

Summary

A reflection on the very idea of nature and the imagination of the wild that it underpins. Starting with the practice of leisure activities in protected sites, or those in the process of becoming so, we endeavour to decipher an often ambiguous relationship with nature. The nature wished for is not natural and natural nature is not wished for (Krieger S-J et al., Loisir et Société/Soc Leisure 40:25–38, 2017). On our terrain, while recreational users claim to be seeking fusion with the wilderness, their search for what is aesthetically pleasing and their need of access takes them back to a nature that has been "managed" but that corresponds to how they imagine re-wilding. It remains to be seen how urban areas and their peripheries can meet this need for mythical wildness.

Keywords

Recreational uses · Need for nature · Mythical wildness · Relation to the urban · International comparison

Introduction

"*Because man's being is made of such strange stuff as to be partly akin to nature and partly not, at once natural and extra natural, a kind of ontological centaur, half immersed in nature, half transcending it.*" (O. y Gasset, 1941, Historia Como Systema. Translated by H. Weyl, E. Clark, and W. Atkinson). This twin anchorage of the human appears all the more clearly, in our opinion, in the recreational user of nature, an avatar of the centaur figure. In this respect, outdoor leisure activities are an important key to understanding our relationship with the environment. The purpose here is to examine what is meant by nature: how can we define it? Are the elements of nature found in an urban context actually "nature"? How do the users relate to nature? Does the nature to be found in towns fulfil the need for nature?

To answer these questions, two distinct but comparable terrains were selected: the Saguenay—Saint-Laurent Marine Park in Canada, more precisely in Quebec, and what has become the Marine Natural Park of Estuaire de la Gironde et de la mer des Pertuis in France (Aquitaine). This is, therefore, a Western, international comparison between France and Quebec, whose policies for the protection of nature tend towards a similar management approach. While there are many ideas circulating about conceptions and relations to the environment of both sides, and they regularly inspire environmental challenges in terms of management, few comparative studies have been produced up until now.

Our field of study is, in France, the south bank of the estuary of the Gironde from Macau to Soulac-sur-Mer and, in Canada, the north bank of the Saint-Laurent estuary, from St-Fidèle to Les Escoumins. Despite their diversity, as for their respective dimensions and configurations, these estuaries are defined as natural, even wild areas, with identified environmental issues (water quality, population and state of health of flora and fauna, fossil fuel, etc.). Both are subject to protective measures (species declared endangered, national park or Natura 2000,[1] Marine Park established or being created during our survey). In addition, there are large numbers of leisure users, on land and sea.

The habitual user at the centre of this survey has been defined, for the purposes of our research, as a user whose

In reference to the article by B. Kalaora (2001).

S.-J. Krieger (✉)
Irstea, Bordeaux, France

[1] The Natura 2000 network, reconciling nature conservation and socio-economic concerns, is a set of natural sites in Europe, on land and sea, identified for the rarity or fragility of the wild species of flora or fauna and their habitats (cf. official website of the ministry in charge of the environment: http://www.developpement-durable.gouv.fr/-Natura-2000,2414-.html, viewed on 03.06.2015).

C. Machemehl et al. (eds.), *Reclaiming and Rewilding River Cities for Outdoor Recreation*, Estuaries of the World,
https://doi.org/10.1007/978-3-030-48709-6_7

relationship to the estuary can be typified by continuity over time or regularity. The majority of the interviewees come from urban areas and some live in the city, but all visit the area regularly. We interviewed them along the south bank of the Gironde estuary and the north bank of the Saint-Laurent estuary, in natural spaces relatively close to the town in the Medoc (between 25 and 100 km from Bordeaux) but farther away for the North coast (approximately 150 km from Chicoutimi for example) in Canada. Among the numerous outdoor sports practiced on the two estuaries, we chose the following[2]: walking, an activity that has been democratised and is massively practiced, and canoeing, sailing and horse riding, more dependent on sociological variables and offering a range of user profiles and so different points of view on our subjects. Based on different case studies (international natural spaces with differentiated regulations) and specific variables (leisure uses and practice modes) or more conventional sociological variables (sex, age, diploma/occupational category, place of residence and geographical origin), our comparison is mixed within the meaning of M-B. Miles and A-M. Huberman (2003).[3] The cross-referencing of these different variables and the use of the qualitative method contribute detailed comprehension of the relationship leisure users have with nature.

In total, during 6 months of field work in immersion, we carried out fifty semi-structured interviews and nine participatory observations around the Saint-Laurent and forty-four interviews and sixteen participatory observations around the Gironde. Focused on the comprehension of the processes, the pairing of the interviews and observations provides access to different types of discourse (including those that are particularly rich in situation). The triangulation of the information then makes it possible to articulate conducts-practices and discourses-representations-beliefs in order to give a better account of the relations between people and nature. The following analysis is based on the materials obtained in this way, and is illustrated with quotes, typical of a vast corpus of interviews attesting to the importance or recurrence of certain attitudes or feelings.

Examining these leisure users makes it possible to grasp the opposition between urban and natural but also the complementarity at play between them. In the first section, we analyse the different dimensions of the experience of outdoor leisure as lived by the users, including the need for nature. In the second part, we try to understand what environment the leisure users of nature are looking for by trying to grasp the ambivalence of the relationship with nature through an analysis of the myth of the wild.

Leisure Use: Between Re-creation and Recreation?

From the Need for Nature...

Expressed in different ways but always present in the discourse of the people we spoke with, the need for nature is a recurring theme among users, as shown by numerous authors before us (Chamboredon 1985; Kalaora 2001; Massena-Gourc 1994) and as other research confirms again today (Ginelli et al. 2014). Nature is a *"refuge"* for our interviewees, where they like to come and *"take the air"* or even *"recharge their batteries"* during a walk for a few hours or an escapade of a few days: *"And now I'm in Bordeaux, I go into the Medoc countryside to take refuge [...] It's a breath of fresh air and we can put aside the town, work, etc."* (horse rider, student, living in Bordeaux). The need for nature expresses itself in concrete terms as a mental and physical necessity for these users of nature. The canoeists and "sailies"[4] speak of their leisure activities as *"a therapy, a way of letting off steam"* that gives them *"zenitude and plenitude"*. Outdoor leisure activities, therefore, seem to play a role as a palliative for modern life, illustrating the analyses in terms of pleasure of J-M. Le Bot (2013, p. 7) who considers this experience of nature as the "releasing of tension experienced as unpleasant".

With our interviewees, this need for nature is presented more particularly in variations on the theme of water. The need for water is doubly fulfilled in recreational users of the north bank of the Saint-Laurent whose activities can take place in turn on the Saint-Laurent estuary and the Saguenay fjord. But even more so with users in the Medoc region, whose activities vary according to whether they take place around the estuary of the Gironde or the Atlantic ocean. This appeal of water can apparently be explained by experience in their life: *"I was brought up on the edge of the water"* (multiple-leisure user, science diver, living in Tadoussac), which is confirmed by J-M Le Bot (2013, p. 6): *"the experiences, positive or negative, of the environment, more*

[2] This chapter mobilises part of the results regarding the relationship to the city and the question of the urban covered within the scope of a sociology thesis (Krieger 2015) aimed at understanding recreational users' relation to nature.

[3] In the opinion of the authors, the comparison can be made according to two principal modes of investigation: variable analysis and case-based analysis. Whereas the former draws on a large number of cases in which the variables and their relations are analysed, the latter starts with a small number of cases and considers a case or a site as a full and complete entity and only envisages comparison in a second instance. Our approach which crosses cases and variables therefore turns out to be a mixed analysis.

[4] A "saily" is a person who navigates with a sail. Their occupation, type of craft or place of practice are of little importance; they move using a sail, therefore by dint of wind power, which distinguishes them from all the other sailors who use motor engines. This term is used in colloquial speech, but as our interviewees define themselves in this way, we respect the nickname in our writing, placing it in quotes.

or less urban or more or less rural, that we lived through in childhood, condition the relationship we have to this same environment once we have grown up".

While a certain geographic proximity ensures emotional proximity which, for the users concerned, comes from habit, more distant users explain that, on the contrary, it is because they come from outside that they appreciate this environment, which provides a break from their daily routine. Ordinary or extraordinary, beyond a habit from childhood, the two registers are invoked to explain the appeal of nature; nature thus becomes a setting for the practice of their leisure activities. But while the interviewees express themselves in terms of need, nature leisure activities are an experience that covers multiple facets for the enthusiast.[5]

. . . To the Experience of Nature

Contemplating the beauty of nature is the prime objective of outings. Landscapes, colours and variations in light seduce the nature lovers. Water, "vital element" for the canoeists or the "sailies", is also highly appreciated by those who walk or ride horses along the estuaries of the Gironde and the Saint-Laurent. This attraction for the coast is expressed in terms of landscape: "*And so I really love the sea, the water, I think it's the infinity you see that makes me*. . . *There, when I talk about nature I think of a view into the far distance*. . . *Nature, for me, is that, rather. It's the landscape of the mountains in the distance, the sea, not closed,*" (rambler, hairdresser, living at Saint-Yzans-de-Medoc). B. Kalaora (1998, p. 133–134) explained some 20 years ago that "the coastline is less a natural milieu than a scene, a landscape, a pleasant view". The people living along the coastline share this attraction to the interface between land and sea with island dwellers, showing that while it is at the heart of the identity of the latter (Lewis and Devanne 2014), it is not exclusive to them, even though people in the Medoc describe their region as a "peninsula".

This predominant place given to aesthetics in the relation to nature perceived as a work of art dates back to the tourism of England at the end of the eighteenth century (Conan 1985). According to M. Conan, the picturesque is "*the effect of a cultural attitude, the product of a singular way of directing the gaze which allows the spectator to look at nature as if it offered the spectacle of a landscape painting*" (ibid., p. 176). And it is effectively this culture of tourist landscape, defined by beauty, originality and disproportion (Depraz 2008), which marks the relation to nature even today. This holds for our field as well as others (Lewis and Devanne 2014), or even, more generally, on protected coastal sites (Kalaora 1998), although other dimensions have emerged and grown. While some of our interviewees effectively materialise their relationship to nature through the practice of painting or photography during their leisure time, all are influenced by this picturesque way of seeing in appreciating their environment. Not specific to ramblers (Niel and Sirost 2008), nor the appendage of tourists, this picturesque landscape approach is shared by all our interviewees: "*The aesthetic experience did not emerge exclusively from the distanced and culturally informed gaze in a context of breach with the ordinary, but in the tension between, on the one hand, the ordinary, the familiar, the routine and, on the other hand, the extraordinary, the foreign, the unusual*" explain A-S Devanne and S Le Floch (2008, p. 125).

The "natural spectacle" allows us to go beyond a purely visual relation to nature and apprehend the associated sounds and smells, such as birdsong and the call of the whale, the smell of seaweed and mushrooms, or feeling such as "*the cool, light breeze when the tide changes direction*". This contemplation through more than one sense becomes a real incursion into nature. And this is where the importance of the chosen recreational use lies. As A. Niel and O. Sirost (2008, p. 192) explain: "*each group claims to have a monopoly on a "landscape view" and they all maintain it jealously*". The hiker, for example soaks up a territory by changing speed according to whim; the canoeist can reach places that are otherwise inaccessible; the added height on horseback allows the rider to discover new scenery, etc. Recreational use then becomes a medium for the relation to nature, which the user strives to be included through the practice of galloping or various forms of board sports. While technique or equipment can be seen as obstacles to the desire for fusion, as for nudists who free themselves from them, on the contrary they can be considered an extension of the body that brings one closer to nature by playing with felt sensations (Le Bot 2013). It is therefore through these *media* that the leisure user experiences and defines their relationship with nature. Nature becomes the environment in the sense of what surrounds outdoor leisure activities: "*Now nature has several connotations in my mind. Since I started sailing everything has changed. Before, there was no notion of the aquatic in my mind: when someone said nature to me, I saw forest, I never thought of the sea.*" (multi-leisure practitioner, waitress, living in Tadoussac). These reflections illustrate the idea according to which "*by living actively in the space, posing physical closeness to the environment as a principle, the people involved in outdoor sports enter an emotional*

[5] Due to the format chosen here, the chapter focuses on certain aspects of the recreational use of nature related strictly to the relationship to nature. Nonetheless, both the practice of a leisure activity and the frequentation of the natural spaces pose questions about how we relate to other people and how we access nature. For a more in-depth analysis of the various aspects of the recreational use of nature, cf. "Le kayak de mer: du "besoin de nature"... à l' "expérience écologique"" (Krieger and Ginelli 2015).

complicity with the landscape" to use the conclusion of A. Niel and O. Sirost (2008, p. 197).

A Healthy Mind in a Healthy Body

Combining their summertime with a necessary spiritual pilgrimage, some interviewees do not hesitate to compare their outdoor sports practice with a religion: *"I'm going to explain something weird: it's as if it were a religion. Because the Moslem must go to Mecca! (laughter) For us, this here [Bergeronnes], is Mecca! It's our religion, we have to come here! And Kamouraska is the Vatican!"* (canoeist, teacher, living in Montreal). While this quote, as clear as it is provocative regarding the role conferred on outdoor activities, may seem anecdotal, it nonetheless reveals the spiritual dimension of the recreational uses of nature as experienced by some of our contacts who speak in concrete terms of meditational practices. At the same time, the activity itself, through the relation with nature it offers, imposes a form of spirituality when faced with the grandeur and power of the elements with which a sort of communion is sought. This aspect, which B. Kalaora (2001) also mentions, expresses the vitalist approach to nature, one of the dimensions of the transmodern form of recreational practices described by J. Corneloup (2011). The author, in reference to A. Berque (2000) speaks of "*the production of an ecumene that binds the individual strongly with their life milieu*" (Corneloup 2011, p. 9).

While meditation turns out to be a dimension of recreational use, this spirituality can be experienced very actively by recreational users who, in their speech, do not really make a distinction between these meditative, even religious, and sports aspects. The first people involved in outdoor leisure activities, marked by the healthy living trend of the time, developed scientific and moral arguments about the usefulness of this type of activity. This reference to the rationale does not seem to be current today, while recreational users incorporate a spiritual dimension into a sports practice renewing with the philosophy of "a healthy mind in a healthy body". Recreational uses offer a very concrete prophylactic dimension for, in a society marked by the seeking of well-being where physical activity becomes a standard, the important thing to be in good health, to "keep in shape", is to move, as repeated in the advertising spots that some of our contacts do not fail to remind us of.

Ecologised Leisure Activities in Wilded Sites?

From the Development of Natural Spaces. . .

The practice of outdoor recreation expresses a demand for natural spaces. In the 1960s, explains G. Massena-Gourc (1994), "spaces left free and vacant" became "collective urban facilities" for city dwellers in need of nature. From the 1980s, J-C. Chamboredon (1985) saw the French countryside being transformed into nature to become a place that repairs urban life, a place of "disinterested" enjoyment. G. Massena-Gourc (ibid.), a decade later, reaffirms this idea of a modern, urban relationship to nature, therefore understood as different from a productive relation to the countryside, expressed as a desire for landscape and a change of scenery. Since then, nature has been developed to welcome these leisure players. For example, at Port-au-Persil, in Quebec, the pier has recently been redone and toilets installed for the security and comfort of nautical users and, more generally, of the visitors. Among our contacts, while opinions are divided about the appearance of the actual building work, all agree that the development was necessary. In the same way, our contacts declare generally that they appreciate the bicycle paths, rambling paths and picnic tables along the way, but all are afraid of overdevelopment. These recreational users wish their nature activities to be recognised, even legitimised, and are therefore expecting the public authorities to do the necessary to allow them to practice their leisure activities, but at the same time conserving the natural character of these spaces, in line with their imagination.[6]

While aesthetics appears to be an important value accorded to nature, the users appreciate the elements of ordinary nature as well as the nature that is remarkable in the sense of Larrère and Larrère (2009). Evidently, whales are the stars around the Saint-Laurent, but this does not prevent the canoeist from observing starfish and sea birds. Moreover, the kayakers in Quebec say they prefer the Saint-Laurent, or even the fjord, to the lakes because *"there's life all around"* which contributes to the effect of surprise they look forward to when they see a beluga or an orca, for example. The presence of the fauna is, therefore, a part of the wild beauty of nature. According to B. Kalaora (1998, p. 143), "*wildness remains a symbol more than a reality: we require it to satisfy the unconscious, but we are actually looking for pleasant, comfortable nature*". Which is the sense of the following excerpt: *"It's good, it's wild, but at the same time not. . . it's not the jungle but it's, I don't know, not too well*

[6] The imaginary construct of an "unspoilt nature", a term used by some of our contacts to describe a wished-for nature with little interference from man, in the double sense of scarcely developed and scarcely frequented—an interesting aspect on access to nature which should be kept in mind as it cannot be expanded on here.

signposted…, it's not over-supervised, its… but that's what's needed… To my taste, I feel its good, there you go." (rambler, hairdresser, living in Saint-Yzans-de-Medoc). And therein lies all the ambiguity, which is far from new.

… To the Mythical Wilding

B. Kalaora (1998) explained the paradoxes engendered by the contradictory expectations of the public: there is a demand for services and development to favour access to nature and outdoor leisure activities, but at the same time the development must not resemble urbanity so that the wild nature of the place is still credible. And although, still according to the author, users demand wild nature, they, in fact, enjoy a nature that is somewhat managed. In the same way, G. Massena-Gourc (1994) explained that the nature that is wished for is not natural and natural nature is not wished for. This tension, according to B. Kalaora (2001), is amplified by the weight of aesthetics in the definition of wild. Wild is defined by the stakeholders through terms such as composition, harmony, symphony and not disorder, chaos, emptiness (ibid.). The wild, which may thereupon be described as mythical, renews and displaces the human gaze towards horizons that are different from the usual, i.e. freed of urbanity.

On the other hand, as shown by G. Massena-Gourc (1994), certain artefacts are accepted or even appreciated because, as they are part of the history of the place, they merge into the idea the users have of the surrounding landscape: *"Because despite it all, it may well be wild, natural, you take the most remote spot in the area, the estuary, you will stumble on a duck hide or a stilted fishing hut. In spite of everything, there are always, but these things are really deeply embedded in… these are not things that shock"* ("saily", fireman and living in Pessac). In fact, *"the "recognition" of a facility and its importance are not measured solely by its objective footprint on the landscape […], but also the degree of "naturalisation" including the landscape, depending quite broadly on the time it has been there, or when history becomes "nature""* (ibid., p. 295). The author quotes the example of life in the hut that requires *"heavy equipment but that is sung in the Provencal tradition"* (ibid.), of which the hunting "tonne" or the stilted fishing hut are avatars in the Medoc. Whether ploughed or fallow fields, hunting hides or fishing huts, these elements refer to culture, as understood by H. Melin (2011, p. 5) as *"relations with the natural environment, the practices developed in nature, the traditions related to the milieu as well as the empirical knowledge of the local population"*. Clearly, still according to the author, *"culture constitutes nature as much as it originates in nature"* (ibid.) and it is indeed the representation that seems to be shared by the majority of our contacts who have incorporated anthropogenic elements into the natural landscape.

ἔρημος (érème) Versus Oecumene/Urban Versus Rural

Nature is first of all defined in opposition to the town. Town which, irrespective of its objective size—Bordeaux, Montreal, Paris and Quebec are mentioned —is seen as large, frequented by *"the crowd, the masses"*. It then becomes *"oppressive, stressful"*, a far cry from the city of Enlightenment that frees the individual from their traditional bonds as Émile Durkheim described. Even though it is still appreciated by some as a place of culture and consumption, it is above all the disadvantages of the urban that retain the attention: the foul smells and the incessant noise. The car denigrated by our contacts because it is noisy and smelly, and especially because it causes pollution, becomes in their speech like the symbol of the polluted urban environment they are fleeing. The city is supposedly the work of modern man, disconnected and dominating nature, whose tarmacadam and concrete, that cover urban spaces, emphasise the breach: *"Nature is beautiful! I find it raises my spirits to see everything that's beautiful and it smells so good. It's not the city, is it! it's the complete opposite to the town! (laughter) it's exactly the opposite adjectives of everything that is negative about the town!"* (hiker, retired, living in Verdon-sur-Mer).

The negatively connotated artificial image of the town illustrates the research of N. Blanc (2000) on nature in town. The fantasy of controlling or even expelling nature from the urban milieu, as people including Georges Haussmann strove to achieve in France, for reasons of hygiene and comfort of the inhabitants, in the end worked better in the collective imagination than in reality, for the town remains an ecological niche that is home to more than twenty species of animal including hedgehogs and foxes, and amphibians who come and live there "freely" (ibid.)… Elements of nature, they are nonetheless at a far remove from matching the common representations of nature mentioned by our interviewees. In fact, while some acknowledge elements of nature in towns such as a tree or a bird, these are not, for all that, in their opinion, nature: *"Automatically, we say nature, I'll say the forest. That's my daily environment. Nature, everything evokes… here for example I'm not going to say its a village I'm going to say nature […] Nature… I'm never going to talk about the city of Bordeaux as nature. I won't see anything that reminds me of nature. Even if there's a tree I'm not going to think that."* (rambler, estate agent, living in Grayan-et-L'hospital). Nature is assimilated to countryside, detached from its productive dimension which marginalised it (Bontron and Morel-Brochet 2002), the former being defined by the absence of human activities and obligations: *"peace and quiet"*. According to A. Berque (2011), as in previous eras, the rural banished the woodland outside the world, the urban rejected the rural outside the world; the country and the wilderness, therefore, finding themselves on the side of έρημος (érème) the opposite of ecumene, that is to say the town. The urban dwellers have naturalised the countryside according to the words of the author. This dichotomist vision explains the definitions given to the terms country, wild and nature in the speech that tends in the majority to merge to oppose the urban, civilisation or even the artificial.

Conclusion

While the elements of nature in town do not currently match the definition and representations of nature, can they meet the recreational users' needs of nature? The town symbolises modern ill-being and does not seem to be able to fulfil the compensatory need for nature of these users who seek to break with routine.

In search of mythical wilderness, these users flourish in wilded places. The recreational user is a Centaur, a creature of both nature and civilisation, and they seek a game park in their image: nature that is developed and wilded to practice their leisure activities in compliance with their imagination.

After the countryside, it is maybe the turn of the town or at least of the edges of the town to be wilded by the city dwellers to meet their needs for nature. For these places for practicing outdoor sports are the environment of the leisure users, therefore a "socially invested nature" as defined by B. Picon (2012). And so if the rewilded town can conceal its part of artifice and reveal its "wild beauties", it could well seduce the city dwellers seeking a change of scenery, or even become a place of symbolic and identity investment for these recreational users.

References

Berque A (2000) Ecoumène, introduction à l'étude des milieux humains. Belin, Paris

Berque A (2011) Le rural, le sauvage, l'urbain. Etudes rurales 187 (1):51–61

Blanc N (2000) Les animaux et la ville. Odile Jacob, Paris

Bontron J-C, Morel-Brochet A (2002) Tourisme et fonctions récréatives. In : Repenser les campagnes, La Tour d'Aigues, Editions de l'Aube, pp 173–193

Chamboredon J-C (1985) La "naturalisation" de la campagne : une autre manière de cultiver les "simples"? In: Protection de la nature: Histoire et idéologie. De la nature à l'environnement. L'Harmattan, Paris, pp 138–151

Conan M (1985) Découverte et invention du Yellowstone. Esquisse de l'histoire de la création d'une culture visuelle aux Etats-Unis au 19e siècle. In: Protection de la nature: Histoire et idéologie. De la nature à l'environnement. L'Harmattan, Paris, pp 175–192

Corneloup J (2011) La forme transmoderne des pratiques récréatives de nature. Développement durable et territoires. Économie, géographie, politique, droit, sociologie 2:3

Depraz S (2008) Géographie des espaces naturels protégés: genèse, principes et enjeux territoriaux. A. Colin, Paris

Devanne A-S, Le Floch S (2008) L'expérience esthétique de l'environnement : une tension sociopolitique entre l'ordinaire et l'extraordinaire ? Natures Sci Soc 16(2):122–130

Ginelli L, Marquet V, Deldrève V (2014) Bien pratiquer la nature... pour protéger les Calanques ? Ethnol Fr 44(3):525–536

Kalaora B (1998) Au-delà de la nature, l'environnement: l'observation sociale de l'environnement. L'Harmattan, Paris

Kalaora B (2001) À la conquête de la pleine nature. Ethnol Fr 31:591–597

Krieger S-J (2015) Ecologisation d'un centaure? Analyse d'une appropriation différenciée des enjeux environnementaux par les usagers récréatifs de nature, Thèse de doctorat, Bordeaux et Rimouski, p 320

Krieger S-J, Ginelli L (2015, Juin) Le kayak de mer. Du « besoin de nature »... à l'« expérience écologique »? Nature Récréation, 2: 67–80

Krieger S-J, Deldrève V, Lewis N (2017) Écologisation des loisirs de nature, entre domestication et ensauvagement. Loisir et Société/ Society and Leisure 40:25–38

Larrère C, Larrère R (2009) Du "principe de naturalité" à la "gestion de la diversité biologique". In Histoire des parcs nationaux. Comment prendre soin de la nature?, QUAE. Museum national d'histoire naturelle, Paris, pp 205–222

Le Bot J-M (2013) L'expérience subjective de la « nature » : réflexions méthodologiques. Natures Sci Soc 21:45–52

Lewis N, Devanne A-S (2014) Étude concernant une aire marine protégée aux îles de la Madeleine. Volume 2 – Caractérisation de la communauté et identification des attentes et préoccupations relativement à une aire marine protégée aux îles de la Madeleine. Université du Québec à Rimouski, Chaire UNESCO en analyse intégrée des systèmes marins, Ministère du Développement durable, de l'Environnement, de la Faune et des Parcs (MDDEFP) et l'Agence Parcs Canada

Massena-Gourc G (1994) Sur nos besoins de nature : l'exemple du Massif des Calanques. Forêts méditéranéennes XV:289–306

Melin H (2011) La culture, terreau de la nature. Développement durable et territoires [En ligne] 2:2

Miles MB, Huberman AM (2003) Analyse des données qualitatives. De Boeck Supérieur, Paris

Niel A, Sirost O (2008, Janvier–Juin) Pratiques sportives et mises en paysage (Alpes, Calanques marseillaises). Etudes Rurales 181:181–200

Picon B (2012) La sociologie de l'environnement, l'héritage d'un questionnement sur les rapports nature-société. In: Manuel de sociologie de l'environnement. PUL, Québec, pp 17–31

The Darsena di Milano (Italy): 'Restoration' of an Urban Artificial Aquatic Environment Between Citizens' Hopes and Municipal Projects

Laura Verdelli and Noémie Humbert

Summary

Over the last 20–30 years, an impressive number of European cities have undertaken projects on the waterfront spaces to meet the needs of what is called the 'leisure society'. Our hypothesis is that keywords like 'sustainable development', 'participation' and 'nature in the city' are now becoming 'mandatory' in terms of promotion of urban projects but hide a certain instrumentalism of fashionable concepts. This is what we try to demonstrate through the case study of the Darsena di Milano (Italy), an urban port which has been the subject of various conversion projects after its decommissioning. This case, particularly representative of the possible discrepancy between displayed marketing and realised projects, is not an isolated case. In another extent, other project promoters encourage 'soft' or 'eco-friendly' characters, without having nor the means nor the ambition to give a real thickness to this aspect in the spatial realisation of urban projects.

Keywords

Waterfront · Leisure society · Urban marketing · Inhabitants' aspirations

L. Verdelli (✉)
Maître de conférences en Aménagement de l'Espace et Urbanisme, École Polytechnique de l'Université François Rabelais de Tours, Tours, France
e-mail: laura.verdelli@univ-tours.fr

N. Humbert
Ingénieur en Aménagement de l'Espace et Urbanisme, Chargée de développement à la Ruche industrielle, Tours, France

Introduction[1]

Over the last 20–30 years, an impressive number of European cities have undertaken projects to transform the areas where the city meets water[2] (Verdelli and Morucci 2014). Despite an apparent similarity of the objectives put forward, which refers to the reappropriation of spaces with high added value, strategically located and the possibility of transforming them to meet the needs of what is called globally the 'society of leisure',[3] not everybody shares the same ideas. 'A l'heure où l'image, le signe et le loisir tendent à supplanter la production et l'industrie, la question du devenir et du sens donné aux sites industrialo-portuaires, dans le cadre d'une patrimonialisation et d'une 'ludification' des berges de fleuve urbain, reste totale'. ['At a time when image, sign and leisure tend to replace production and industry, the question of the future of industrial-port sites, as part of heritage and in a frame of 'gamification' of the banks of urban rivers, remains open', translation by authors] (Colin 2012). Depending on the cultural, economic and political context, we are witnessing various types of operations, which vary in scale (in terms of dimensions, strategic ambitions and investments), ranging from the ambitious production of wealthy residential districts

[1] This chapter has been proof-read by Ms. Kathleen McCarron, Project Officer at Interreg 2 Seas (Lille, France), and Mr. Hadrien Herrault, PhD scholar in Spatial Planning at University of Tours (France), and Queen's University Belfast (Northern Ireland).

All the documentary material comes from Humbert, N. (2014). Darsena di Milano—Projet spatial et influences d'acteurs urbains (Master thesis, University of Tours, under the supervision of Paola Savoldi and Laura Verdelli).

[2] A very vast literature exists on this subject. A modest typological essay of these different transformations can be found in: Verdelli, L., Morucci, F. (2014). Renew the identity of the city between port culture and leisure. The case of Livorno. Leisure and Society/Society and Leisure, 37, 58–78.

[3] Simplifying the definition of J. Dumazedier (1962) it is often referred to the 'society of leisure' as to a monolithic entity, but in reality, 'leisure' includes at least: everyday recreation, creative hobbies, recreational leisure, sports activities, cultural practices, sociability ...

C. Machemehl et al. (eds.), *Reclaiming and Rewilding River Cities for Outdoor Recreation*, Estuaries of the World,
https://doi.org/10.1007/978-3-030-48709-6_8

in port areas (Rode et al. 2015) to the simple revegetation of riverbanks. These transformations systematically call upon a panel of key expressions, and sometimes even all at the same time: the return of nature to the city (like the French concept of having a 'green and blue network' in all urban areas); the conservation of cultural but also 'natural' heritages (Verdelli 2014); the provision of sports and leisure facilities; thus, giving a new 'sustainable' image of the city.

In a saturated urban space, not only in terms of physical reality but also in terms of activities that make use of the free remaining spaces in a built environment, a new philosophy tends to impose itself. A series of outdoor activities: recreation, fun, sport, culture (especially festive events for the general public), has now moved from the inside to the outside of cities. The frame of reference is one that promotes a 'non-urban' context to appeal to a certain myth of returning to the countryside, to nature, to a quality of life 'of yesteryear'. This seems to respond to a new demand for natural areas by the city dwellers, which remains vague in the eyes of researchers, but which appears obvious from both the political point of view and the demands of the citizens. In this perspective, the presence of water (rivers, lakes, ponds, small lakes . . .) in the immediate vicinity of cities is often exploited, serving as a backdrop for cultural events. 'Ce phénomène a engendré le développement de ce que l'on appelle 'les loisirs de courte durée' [. . .] Aussi, il était nécessaire d'étudier des aménagements de loisirs, proches des centres urbains [. . .] Pour ce type d'aménagement il faut un 'ingrédient' de base: l'eau! ' ["This phenomenon has led to the development of what is called 'short-term leisure' [. . .] Also, it was necessary to study leisure facilities, close to urban centres [. . .] This type of development requires a basic 'ingredient': water!", translation by authors] (Cazes et al. 1995).

A series of common variables are applicable in a good number of examples where the city–water interface represents a meta-territory, listed in its various portions and components. The boundaries of this vast meta-territory are composed of a certain number of preliminary constants. Among the structuring ones, we will retain the watercourse, as an axis in its material reality and in its mythical value, linked to imagination and fascination, to the construction of social and economic relations (Marié and Tamisier 1982; Marié 1983; Drain 1998; Béthemont 1999); the notions of heritage and landscape in their tangible and intangible aspects; as well as a number of common features:

- Proximity to a dense urban space, often saturated from several points of view (habitat as well as recreational activities).
- Increasing urban demand for 'nature'.
- Growing tourist and excursion market.
- Process of identification, protection and enhancement of cultural landscapes (Verdelli 2008).
- Landscapes considered exceptional that are undergoing (or actively participating in) a process of heritage development.
- Fragile and highly sensitive ecosystems.

We make the hypothesis that a certain number of keywords are now becoming 'mandatory' in terms of promotion of urban projects (sustainable development, participation, nature in the city, . . .) but behind these discourses is hidden a certain instrumentalism of fashionable concepts in a context of strong interurban competition. We try to demonstrate this through the case study of an urban port which has been the subject of various conversion projects after its decommissioning. These projects, very similar from the point of view of the purely spatial elements that characterise the effective transformation of space, have nevertheless been promoted in very different ways, in order to adapt to the evolution of 'mandatory' terms in urban marketing.

Darsena: A Case Study Torn Between the Different Wishes of the Population and Those of the Municipal Council

We will focus here on the history of a very specific example, that of the former port of Milan (Italy) where the evolution of 'trendy' impacts the urban project and its objectives, and where the institutional communication invokes participatory democracy as long as it does not conflict with political and economic imperatives.

Darsena (artificial water basin used for mooring and berthing of boats) is the name given to the ancient port of Milan within the system of artificial canals that connect the city to the regional water network. Milan, while considered as being 'in the middle of many waters', is not on a major river. The city, founded on a system of small rivers—Lambro, Olona and Seveso—was equipped, from the twelfth century, with a system of artificial canals connecting the town to the rivers Ticino (to the West) and Adda (to the East), and, accordingly, to the Lakes Maggiore and Como (to the North) and to the river Po (to the South) for irrigation and transport purposes. The Darsena represents today an area of urban water space whose destiny has become unique. For unknown and unexpected reasons, this site has attracted the affection and attention of the Milanese, to the point of becoming a hot topic of interest in the city. It is a catalyst for citizens, media and public action, it is a rich example for the study of the decision-making process, the organisation of the institutional promotion of urban projects and the citizens' concern about the city–water interface.

Three elements, in particular, aroused our interest and our questions. The first being a sudden political interest in the

2000s, after a period of neglect of more than 20 years. The second being the unusual enthusiasm of citizens and media for this area, all with different demands. Then finally, the third point, despite all the ideas put forward during the last 15 years, the huge gap that exists between the hopes of the citizens and the actual spatial transformation, that remains very modest (in aspirations, programme and physical features).

Research Methods

We propose to focus on both the observation of the process of engagement of stakeholders in the debate and the observation of the modes of action/interaction between the fashionable keywords used in the promotion and marketing and the values supported by official and 'alternative' projects.

The hypothesis is that the characteristics of space result in a multiplicity of interests whose equilibrium has been disrupted by public intervention, the Municipality first intention to transform this space into a car park, provoked citizens and media involvement, promoting instead the respect of the existing heritage, of the biodiversity, of the water, spaces of leisure and nature presence in town. 'L'objectif est de comprendre comment, dans un contexte global longtemps caractérisé par le déclassement de la fonction de transport des sites portuaires, les questions d'aménagement et d'usage encouragent aujourd'hui le développement d'une nouvelle relation entre ville et sites portuaires' ['The aim is to understand how, in a global context long characterised by the decommissioning of the transport function of port sites, spatial planning and uses issues are now encouraging the development of a new relationship between city and port sites', translation by authors] (Paffoni 2012).

We collected written and oral documents with the primary aim of knowing and understanding the history of the place and the debate from the beginning of the 2000s; then capturing its origins and key stages, and then classifying the data in order to compare the arguments for and against. To do this, we relied on the press, blogs, social networks and a series of interviews with the main local associations.[4] Our reconstruction of the course of the events, not being the purpose of this chapter, is, however, quickly sketched since it is necessary to understand the particularly intricate context of the action.

A Neglected Historical Port Which Suddenly Everybody Is Talking About

The site of the Darsena represents both a mythical symbol of Milan's glorious past and a space free of buildings, i.e. an invaluable land opportunity due to its scarcity in the city centre.

A Unique Place, to Defend and Protect

The site of the Darsena is an urban space perceived in a very variable way according to the different points of view, its characteristics conferring to its multiple facets with stakes. Thus, some interested parties defend its character of entrance of the city. Because of its geographical position and its history, the Porta Ticinese has always been a gateway to the city. Although port activity has long since disappeared, the Navigli district still remains a connection between the inside and the outside of the city, the canals and their banks providing rapid access to the countryside of Abbiategrasso or Morimondo. Today, the circuit Naviglio Grande—Darsena—Naviglio Pavese remains the only part of the city where water continues to flow above ground, from the river Ticino to the river Po. This unique character that the canals confer to this district is exploited for **tourist navigation**. Itineraries are indeed proposed by associations to **discover the artistic cultural elements** along the banks of Milanese canals. Then, while **rowing** and other boating activities remain in vogue on the Navigli Grande (promoted, for example by the historic clubs Canottieri Olona and Canottieri Milano, founded around 1890) and Pavese, their banks host **cycle routes**. The association Darsena Pioniera speaks of a 'hinge' between the city centre and the periphery. The central points then defended are the **visibility and accessibility** of this important and totally unexpected urban space in a city where all the presence of water has been systematically obscured or even erased.

[4] *Websites* (September 2013–April 2015): Associations' websites: **Navigli Live—PAN** www.naviglilive.it; **Darsena Pioniera** http://darsenapioniera.wordpress.com/; **Riaprire i Navigli** www.riaprireinavigli.it/; **Amici dei Navigli** www.amicideinavigli.it/; **Comitato dei Navigli** http://www.salviamoinavigli.it/; **Italia Nostra** www.italianostra.org/; **CIVES** http://www.cives.partecipami.it/content/view/10. Facebook pages of the associations: https://www.facebook.com/comitati.navigli; https://www.facebook.com/pages/Darsena-Pioniera/120202078010362; https://www.facebook.com/pages/Associazione-Amici-dei-Navigli/102406353164716; https://www.facebook.com/pages/CIVES-Associazione-Nazionale-Onlus/300145683332262; https://www.facebook.com/pages/Riaprire-i-Navigli-a-Milano/308830319238778; https://it-it.facebook.com/comitati.navigli/; Media websites: www.arcipelagomilano.org; www.ilgiornale.it/; www.ilsole24ore.com/; www.repubblica.it/; www.corriere.it/; www.ilgiorno.it/; www.lepetitjournal.com; www.urbanfile.it/. Universal Exhibition 2015 website: http://www.expo2015.org/it

Interviews (conducted between September 2013 and April 2015): Associations: Comitato Navigli; Riaprire i Navigli; CIVES; Bei Navigli; Darsena Pioniera. Press: Arcipelago Milano; Il sole 24 ore. Contractors: Engineer of the Villoresi Company. Public institutions: Technical Services of the Municipality of Milan; Services in charge of the urban planning of the Municipality of Milan.

For other interested parties, such as the association Amici dei Navigli, it is the issue of **heritage protection** that dominates. The site of Darsena abounds in particular with an undeniable architectural and urban heritage: the entire hydraulic system of the canal network, Milan's historic role as a port, the foundations of the Spanish bastions, the first wooden lock Santa Maria… The Regional law of the sixth of April 1995 is well aware of the difficulties of enhancing this urban sector in relation to the constraints of heritage preservation. Thus, to modify the current image of the area, it is necessary to have a single landscape plan, able to safeguard the integrity of the area. For some, like the association Navigli Live—PAN, it is the tourism's issues that are to be defended. The area being full of elements of interest for visitors, in particular, a unique urban and cultural heritage, the redevelopment of the Darsena should help to develop navigation and tourism activities.

The association Darsena Pioniera [Pioneer Darsena] is committed to defend the **natural resources** that have developed spontaneously (the association takes its name from the 'pioneering' plants that have colonised the site) in the decommissioned space. The association argues that the variety of animal and plant biodiversity, the richness of the landscape and the presence of water thus gathered in a unique way on the same urban site must be preserved for ecological reasons and for the conservation of the quality of the living environment. 'Le domaine de l'eau illustre bien la territorialisation de l'action environnementale en ville, et ce faisant les adaptations urbaines qui ont eu lieu au contact d'une demande croissante d'amélioration du cadre de vie' ['The field of water illustrates well the territorialisation of environmental actions in the cities, and in doing so the urban adaptations that took place in contact with a growing demand to improve the living environment', translation by authors] (Gourlot et al. 2009).

The last issues to be defended are those of the attractiveness and accessibility of the place as to allow **commercial and economic activities**, connected to the intense and lively animation associated with the **artistic, cultural and recreational nightlife**, to remain flourishing.

A Potential of Imagination

Cut off from its role as an urban port since the 1980s, the Darsena is in gradual decline. In the 2000s, this urban space is clearly abandoned, as described by Le Petit Journal: '*60,000 m*^{2} *of a fenced perimeter that have turned into both an open dump and a secret place for stolen goods and dealers of the area*' [translation by authors] (Mais qu'est-il advenu de la Darsena, Le Petit Journal, 29.03.2015).

However, in a way, the fact that the institutions did not feel concerned about the future of this space made it possible for anyone to hope, dream and speculate on its potential future. The lack of interest by the Municipality allowed individuals to imagine, create, virtually detain the ownership of the space. Sentimental and nostalgic people were reinventing a return to the port of the past; artists and dreamers were imagining a new way to belong to this unique place and were inspired by it; while ambitious speculators were speculating on future investments to be considered; all making instrumental use of the recent examples and concepts that have guided the plethora of redesign interventions on European waterfronts.

The 2015 Universal Exhibition 'Feeding the Planet, Energy for Life' and the Project of Spontaneous Garden Oasis: An Unbalanced Confrontation

It was in the year 2001 that the Municipal Council of Mayor Albertini launched the tender for the implementation of the project of an underground car park under the Darsena. In 2003, the first version of the project received a favourable opinion at an official conference, while at the same time, during the autumn, the space was cleared and dried up to launch the mandatory archaeological excavations. The first drafts of an underground car park project were signed by the Municipality at the beginning of August 2003. A framework agreement with the Lombardy region defined the functional and environmental recovery of the Milan canals system and the urban redevelopment of the Darsena through an international design competition (attributed to the project of the team of the French architect J.-F. Bodin). Citizens' actions intensified in 2005: sensitisation meetings, petitions, legal recourse to the TAR (regional administrative court). The period from 2004 to 2006 was marked by numerous modifications made to the initial project, both to meet citizens' and administrative demands and to adapt to the presence of historical remains discovered during the excavation phases. During the relay period between the municipalities of Mr. Carlo Albertini (Forza Italia, 1997–2006) and Ms. Letizia Moratti (Forza Italia/PDL, 2006–2011), the project remained 'frozen'. For more than 6 years the underground parking project under the Darsena was being pushed back, leaving in its wake an abandoned urban space and a heavy legal battle between the Municipal Council and the consortium Darsena Spa; the first claiming the revocation of the concession and the others the damages for not being able to build. In the spring of 2010, the legal battle between the City Hall of Milan and the consortium of companies, is marked by a decisive step that is the return of the ownership of the wharves to the Municipality (April 22, 2010). The competition of 2004 became once more topical, even if, being based on the construction of the underground car park, it must be heavily adapted.

After the above-mentioned fluctuations, that began in 2001 with the project of the parking (pending the completion of which the site of the future construction site is officially closed to the public), it is in 2006, when the Municipality of Letizia Moratti committed itself to support Milan's bid to host the Universal Exhibition, that the strategic interest of Darsena clearly emerged. 'Le processus de réaffectation des espaces portuaires [. . .] est fréquemment déclenché par l'organisation d'une manifestation de portée internationale comme les Jeux olympiques à Barcelone (1992), l'Exposition internationale Christophe Colomb à Gênes (1992) ou l'Exposition universelle à Séville (1992)' ['The process of reallocation of port areas [. . .] is frequently triggered by the organisation of an international event such as the Olympic Games in Barcelona (1992), the Columbus International Exhibition in Genoa (1992) or the Universal Exhibition in Seville (1992)', translation by authors] (Lechner 2006). The theme chosen for the Milanese exhibition being: 'Feeding the Planet, Energy for Life', the question of water resources is particularly evoked, the bid therefore highlights the canals district and its potential development by 2015. The 'Navigli' district (to which the Darsena belongs in terms of location, even if not directly invested by the nightlife activities) is the famous place of Milan's nightlife since at least the 1980s. In 2008, Milan is proclaimed the official host city of the 2015 Universal Exhibition (EXPO 2015).

In the spring of 2010, a wave of citizen actions takes shape in the hope of revaluing the space in perdition and anticipating the institutional actions linked to the mega event which aim more at the theatricalisation of a landscape-panorama type postcard to admire rather to the creation of a landscape to live. The aesthetic emotion related to contemplation seems to be the only form of appropriation anticipated by the municipal project. Clean-up operations of the Darsena (entirely carried out by volunteers) take then place, but doubts hang over the future of the site. Citizens' actions, petitions against the return of an underground car park project and proposals for associative projects, were developed in parallel. In any case, controversies and civic engagement around the Darsena are constantly fuelled by associations and private spheres, including media. The interests for this water space, abandoned by public authorities (and by all more generally, while remaining in the collective memory) during more than two decades, are now reactivated. At a time when citizens' mobilisation is such that all projects are blocked, the association Darsena Pioniera[5] was created with the aim to defend the recovered biodiversity (the site having meanwhile been recolonised by spontaneous vegetation) and the natural space conducive to recreational practices (photography, walking, promenade, relaxation, jogging, water activities and sports . . .).

Darsena Pioniera succeeded in having the Municipality implement a minimal project capable of (re)opening the site to the public. The hasty redevelopment of the space (some trees and shrubs, tourist signs on fauna and flora, wooden bridge . . .), while pursuing a goal close to the restoration of the environment, appears as an inexpensive intervention to public authorities (Figs. 1, 2 and 3).

At the new municipal elections, the candidate who opposes the outgoing Mayor Ms. Moratti (centre-right wing) wins. The Municipal Council, therefore, passes under the direction of the Mayor of the centre-left wing, Mr. Giuliano Pisapia, on 28 May 2011. This historic event marks a real hope of change for the management of the Darsena site and the associations are boosted to build new project proposals.

Citizens' Project Proposals, with Conflicting Objectives

Sensitive Relations Within Civil Society

Each subgroup of local civil society produced its own ideology and objectives, implying divergent interests. This generated a subsequent conflict situation internally to the opposition. The conflict of interest between the renewal of navigation, strongly desired by the District Committee and the association Navigli Live—PAN, and the preservation of the green oasis spontaneously developed in the centre of the basin, defended by the Darsena Pioniera is a good example. '[Les fronts de fleuve urbains] deviennent ainsi une nouvelle destination urbaine, des espaces de convivialité, d'échanges, de rencontres... mais aussi des espaces disputés par différentes fonctions souvent antinomiques' ['[The urban riverfronts] thus become a new urban destination, spaces for conviviality, exchanges, encounters . . . but also spaces disputed by various functions that are often antithetical', translation by authors] (Gravari-Barbas 2004).

However, this situation of conflict of interest did not prevent these same associations from cooperating according to the circumstances and the 'times' of the project. Indeed, Darsena Pioniera and the Neighbourhood Committee, who

[5] The association Darsena Pioniera is very recent. It was created in 2010, by a group of 8 people (gathered around a photographer passionate about fauna and flora). The members of this association have in common the distinction of being professionals who have received a higher education and the desire to do something 'practical'. The fact that the founding group comes from academia and liberal professions leads to a judgment of value by the general public and the media that categorise the association as 'elitist'. Some pronouncements in favour of the defence of biodiversity also leads to the designation, in its pejorative sense, of the activities proposed by the association of 'ecological militancy' ...

Fig. 1 Self-portrait, on the banks of the Darsena before its restructuring, members of Darsena Pioniera wearing as a hat a pot containing a sample of plant species installed in the space of Darsena © Photo Alessandra Mauri

seem to have been two major stakeholders in the debate concerning the requalification of the Darsena site, effectively worked together for the site clean-up operations in the years 2010s.

For the District Committee, the Darsena is a symbolic place of the history and heritage of the city whose decline reflects that of the deterioration of the quality of life in the neighbourhood.

For Darsena Pioniera, the old port is a place of interest in the city due to its geographical location at the gates of the centre and at the same time very accessible via the banks of the canals. This place has become an attraction, with an exceptional beauty and atmosphere. However, its state and uses in the 2010s are unacceptable: 'abandoned space', 'catastrophic state of degradation', 'closed to the use of citizens, which promotes the accumulation of waste' [translation by authors] (http://darsenapioniera.wordpress.com and interviews to Alessandra Mauri and Francesca Oggionni). Nevertheless, for the members of the association, the site is full of unexpected richness that are, in particular, the archaeological discoveries and the spontaneous garden. The latter one is composed of a varied and luxuriant vegetation, which has developed naturally in the zone of the construction site, abandoned since 2006, and which shelters a very original fauna for a city centre.

Different Methods of Action

For these two local associative groups, citizens' consultation is an important part of the process. The Neighbourhood Committee is made up of volunteer residents and aims to defend the quality of life of the area, its role as representative implies that its choices are representative of the wishes of the inhabitants. The use of questionnaires and public communications, as well as public events (e.g. the candlelight vigil on 27 June 2003 called 'Let's lit the quays of our lights') are designed to gather users' opinions. The Committee's website and Facebook page allow its activities to be monitored. The association Darsena Pioniera proposes collective action operations such as 'cleaning and gardening guerrillas' on the quays in the spring of 2009 (with an explicit reference to the 'guerrilla gardens') and events such as the public launch of the cuttings programme, communicating a message of respect for nature, or the operation 'a Hug for the quay' on Sunday, 18 April 2010. But, above all, it effectively used the

Fig. 2 First proposal of the group Darsena Pioniera (2009), left—pre-project picture (the extent of the spontaneous flora is clearly visible)/right—computer simulation of the project © Alessandra Mauri

Fig. 3 Realisation of the first project defended by Darsena Pioniera (2010) © Photo Alessandra Mauri

power of influence of e-participation, with the setting up of a regularly updated website, some blogs and a Facebook page.

The Implementation of the Project by the Public Authorities

The Catalytic Effect of the State of Emergency

Mr. Pisapia's new Municipality, on which the left-wing electorate had big expectations and which is supposedly promoting values such as participatory democracy, nevertheless inherited the management of the works for the EXPO 2015. The urgency of the project only a few years before the deadline would precipitate decision-making. As of July 2012, without paying attention to alternative citizen proposals, the Municipal Council approved a readjusted version of the project that had won the 2004 international competition (Bodin project) to meet the needs of EXPO 2015 and its concept of 'Water City' (the part of the project that consisted of three levels of underground parking under the water basin simply disappeared). The Darsena area is not at the heart of the project, but is part of the overall coherence and coordinated image of the city that EXPO 2015's marketing highlights. The challenge in this type of space obviously remains to produce 'temporary' spaces that can be used as sets for events and become, at the end of the mega event, 'sustainable' spaces.

The numerous projects submitted to the competition (ten groups of finalists are selected during the first selection phase out of the initial 54 applications) use vegetation and materials to create a new identity for the site, or, on the contrary, to redevelop the old one. The urban river landscape dimension becomes the structure of the projects and calls for the re-naturalisation and landscaping of the river (Romain 2010). All projects strive to give more space to pedestrians and cyclists, but the pedestrian connections do not all take the same shape and do not all take place in the same spaces. Some projects exploit the existing symmetry, while others jostle it. Some leave a lot of space for cars and public transports, while others see it as an opportunity to create pedestrian spaces... Most use existing height variations to separate low and high walks. According to the press, it seems that J.-F. Bodin's team most impressed the jury for its understanding of the historical geomorphology of the area and for its sensitive approach, and that the project would be able to add significant value to the use of the space, matching both the habits and the expectations of the current local community.

Only the emergency situation, caused by the deadline of the EXPO and the fear of losing the funding granted in this context for the restoration of the waterways, imposed the exit of the dead end, with a return to action, even if it meant drastically changing the slogan: from the car park to 'L'acqua, l'elemento naturale di Milano' ['Water, the natural element of Milan', translation by authors] (exhibition panels bordering the site, 2014), and having to put aside any form of participation. The final version of the project is estimated at 17,000,000 euros, including VAT (the 2004 competition foresaw a budget of 20 million euros). On 18 January 2013, the general manager of the EXPO 2015 definitively approves this choice for the final design of the project of renovation and redevelopment of the Darsena.

In emergency situations, the time pressure is such that it forces us to react in the moment. 'L'urgence constitue une sorte de 'cas limite de l'agir', parce qu'elle bouleverse le rythme temporel habituel de l'action' ['Urgency is a kind of 'borderline' for action because it disrupts the usual temporal rhythm of action', translation by authors] (Pierron 2006). The impossibility of delaying or postponing actions implies 'une crise de l'initiative: la décision d'agir ne peut plus être garantie par une délibération lente et mûrie' ['a crisis of initiative: the decision to act can no longer be guaranteed by a slow and mature deliberation', translation by authors] (Pierron 2006). In an emergency, remaining undecided would appear as an 'aveu d'impuissance' ['admission of powerlessness', translation by authors] (Pierron 2006). However, these generalities seem to be entirely consistent with the situation of the Milanese political decision makers in the case of Darsena. In these emergency conditions, due to the risk of not being able to benefit from the subsidies specifically granted for the redevelopment of the old port and the possible failure of part of the EXPO operation, the Municipality of Milan closed all discussions with citizens.

As it is possible to make clear through the commitment of the association Darsena Pioniera, reconstituted from the information available on the association's website. On 14 June 2011, the Milan Municipal Council organised a consultative referendum in 5 parts, the question no. 5 being: '*Do you want the Municipality of Milan to reorganise the Darsena as a port of the city and as an ecological zone and work progressively for the hydraulic and landscape recovery of the Navigli system on the basis of a specific road map of feasibility?*' [translation by authors]. Following the results (out of the 50% of eligible voters who went to the poll, 94.32% voted favourably) the association Darsena Pioniera restarts active resistance by sending its new project proposal called 'The island of the quay' to the Municipality; by participating in various events, including the public meeting 'Rethinking the Dock' with CIVES (19 November 2011), and the 'Partecip-AZIONE!' (17 December 2011); by a new campaign of mailing to the Municipal Council; and, finally, by sending the new project to the press. In July 2012, the Municipal Council approved the modifications to the winning project (Bodin) of the 2004 competition for the redevelopment of the wharves, without having reacted in any way to citizens' requests. And, in particular, without having taken into consideration the new proposal from Darsena Pioniera, yet compatible with the Bodin project and supported by professional expertise. Months of activism will lead, on 03 December 2012, to a meeting with Mr. Gianni Confalonieri, Director of the Institutional Relations of the Municipality, who will reaffirm the need for the city to comply with the EXPO project schedule and the 'no' to maintain the oasis, with the argument that taking into account this variant would require too much time. On 23 January 2013, a municipal meeting communicates to the citizens the changes that will imply the realisation of the project Bodin revisited. 06 February 2013, marks the end of all hope of being heard for the association Darsena Pioniera, since the service of green spaces of Milan completely cleared out the area of the Darsena (thus removing the very object of the fights of the association).

The Failure to Take into Account the Dreams of the Inhabitants

These somewhat hasty steps, put in place by the Municipality for the requalification of the site did not leave any room for public debate and destroyed any possibility of imagination. 06 February 2013, marked the end of citizens' expectations: the site of the Darsena is totally cleaned and cleared by the city services to begin the redevelopment of the quays for the EXPO 2015. The utopias built by everyone on the potential future of this space must be abandoned given the Municipality's determination. The hopes of seeing the Darsena once again become the harbour of the past, with the reopening of the canals, the hopes that it will become a passenger transport pier (workers/tourists), the hopes that it will become a natural park (the project of spontaneous garden oasis), ..., have been dashed. The historic dike, once consolidated and restored, will remain the dominant feature of the North Shore, offering walks, leisure, shopping, recreational activities, events ... On the South Shore, port operations (towing, mooring and boat maintenance) will take place. The 2004 project will be significantly simplified: one of the two pedestrian bridges planned by the Bodin team will not be realised, as the wooden pontoons and small kiosks and the botanical garden (which was to include an aquatic part) that will leave the place to a simple garden shaded by trees. The opening of the canal towards the Varenna basin will not be carried out, but a bar/cafeteria will be proposed along the northern bank. For the Neighbourhood Committee, the long-awaited cultural approach for the requalification of the Darsena has been forgotten in favour of a 'totally commercial' project.

The Reversal of the Municipal Council Policy

The emergency situation enabled the municipal team to break the deadlock and bring the project to execution. Therefore, as the philosophy of action argues, a state of emergency reverses priorities and, in a case such as that of the Darsena,

Fig. 4 The project completed, on the eve of the opening © Photo Paola Savoldi

allows projects that were hitherto pending or in indecision, to finally return to the forefront and move on to implementation.

Having to integrate the general context of the 2015 World Expo, the final (and only realised) project was born from an adaptation of the Bodin project to the overall programme of the 'Waterways'. It will have to become part of the new 'blue and green' ring around Milan. For the 'Vie d'acqua e Darsena ritrovata' ['Waterways and Darsena rediscovered', translation by authors] programme consists of a 'series of initiatives to improve the landscape and the environment of the open spaces in the western belt of the city, the canals and the network irrigation' [translation by authors] (www.expo2015.org, accessed on 02.03.2014); the Darsena basin is thus reorganised as a navigable water body.

The official policy of the Municipal Council thus adapts to popular opinions, with a total inversion in the meaning of the keywords and a realisation that has changed almost nothing to what originally planned, the reshaping concerning more the language than the actions: the reappropriation of this space by citizens, the new green spaces (Milan being one of the big European urban areas where the lack of public green spaces has been considered as dramatic at least up to the beginning of the 2000s—Ambiente Italia 2006), a rediscovered water body, the revival of the memory of the port and its heritage, 'the Darsena will be returned to the city and will become a historic place and a symbol of Milan. The project involves the restyle of the old port and the redefinition of adjacent spaces. The intervention focuses on the renewal of the banks, with new walking areas and a marina for tourism' [translation by authors] (www.expo2015.org, accessed on 02.03.2014).

At the same time, the objectives of the various groups and associations that are difficult to reconcile are also decided upon: the project does not provide or guide any specific practice and may be limited to prohibiting those that are considered undesirable (the practice of skateboard for example). In any case, any effective practice on the site remains to be built, the Darsena having been inaccessible during the three decades that followed the cessation of historical navigation. The Figs. 4, 5, 6 and 7 show an essential mineral and smoothed space where the street furniture is austere and where the display of volumes in space (which could have been based on geomorphology) is simplified to the maximum, proposing a refined aesthetics.

Fig. 5 © Photo Paola Savoldi

Fig. 6 © Photo Paola Savoldi

Fig. 7 © Photo Paola Savoldi

Conclusion

If some of the contemporary urban projects can be considered as eco-friendly, a greater number just call themselves as such. This means that among the projects presented under the banner of sustainable development, promoted as being harmoniously integrated into their environment, having been conducted with the participation of users, where vegetation and green spaces take an important place ... some may ultimately be quite different. This is the case of the Darsena in Milan, which, perhaps because of a very complex situation, shows a striking mismatch between the political discourse that accompanied the project process and the resulting transformation of space.

The dead end in which the future of the site was in, had been overcome with the urgency of concretising a project before the beginning of the EXPO 2015. To disguise the imposition of a project that appears as very mineral, built without consultation with a population which had, however, manifested itself massively to be able to participate in the future of the site, the Municipality used keywords impossible to circumvent today, perhaps overused but certainly seductive and federating, those built around: sustainable development, the place of nature in the city and the culture of leisure. In fact, the recreational, leisure and sporting practices developed more and more spontaneously since the 1990s throughout Europe around urban areas where water is present, seem to play an important role in the reconfiguration of cities. Without, however, being systematically at the origin or at the heart of urban redevelopment projects, they constitute an unavoidable challenge.

However, recreation and leisure seem to be, in our case, an interest to defend like many others, without being at the origin of urban transformation. On the contrary, the organisational interactions between urban interested parties are the source of procrastination and strong positions that will fuel the debates and lead to the definition of the territorial transformation project. The idea of not being passive and not accepting to be imposed on the project triggered a real citizen reaction in one of the cities where the inhabitants are the least involved in Italy. While the redevelopment of riverbanks and wharves throughout Europe rely on a rediscovered naturality, the reconfigured site of the Darsena surprises by its minerality. This, despite the media green varnish affixed to the project: '*The sun of the days after Easter [2015] has highlighted the lasting charm of this area, formerly a port, whose development seems to be entirely consistent with the theme of the environmental sustainability of the EXPO 2015*' [translation by authors] (Lavori in Darsena al capolinea: il 'porto di Milano' torna a vivere per Expo, Il Giorno, 7.4.2015).

Through this observation of the process which led to the functional and symbolic reuse of a part of the city to which the collective imagination attribute values and desires (Vermeersch 1998), we were able to show, with the case of the Darsena of Milan, that it is more the upheavals of balance (multiple interests associated with space: inaction, state of emergency on decision-making and hasty action) which defined the reconfiguration of the city on water, rather than the projection on possible practices. The example of the Darsena also highlights the tensions between leisure activities, river cities and urban issues, since the Milanese canals district is a lively living area with high gentrification potential (Gralepois et al. 2018), but the nightlife and leisure activities that develop around the aquatic elements cause urban management problems (traffic, parking and safety). The Municipality of Milan is exploiting as much as possible the opportunity offered by the EXPO 2015 to promote a territorial marketing that could be called 'green', to boost the local economy. The opportunity to (re)build a positive image through the enhancement of the 'blue' network in the urban landscape (Bonin 2007) of a city that suffers a strong deficit in terms of image of 'city where it is good to live', passes by the capitalisation of the potential generated by the mega event and the associated media promotion.

The municipal project reuses in an instrumental way all the keywords and arguments mobilised by the citizen projects because they (1) correspond well to the contemporary goals pursued by the EXPO 2015; (2) allow the display of the project as 'a global urban project'; (3) appear more adapted to the circumstances than the image, which supported the first Bodin project associated with the car park; and, finally, (4) produce a publicly displayed representation aimed at convincing citizens and associations as well as inhabitants that their voice was heard.

The case of the Darsena, particularly representative of the possible gap between what is told and what is done, illustrating that behind common words can hide different meanings, is not an isolated case. In different ways and elsewhere, other project leaders promote 'soft' or 'eco-friendly' characters, without having the means nor the ambition to give real depth to those aspects beyond the fashionable descriptions that accompany the spatial realisation of urban projects.

References

Albertini G, Lomartire CM (2011) Dans le bureau du maire - Neuf ans à la direction d'une métropole qui change. Guardamagna Editori, Varzi

Ambiente Italia (2006) Ecosistema urbano Europa. Dexia and Legambiente, p 87

Bernoux P (2014) La sociologie des organisations: initiation théorique suivie de douze cas pratiques. Seuil, Paris

Bernoux P, Bion J-Y, Cohen J-C (2001) L'analyse des systèmes d'acteurs. Cahier n°1: Diagnostics de territoires et systèmes d'acteurs. Certu, Lyon

Béthemont J (1999) Les grands fleuves, entre nature et société. Armand Colin, Paris

Bonin S (2007) Fleuves en ville: enjeux écologiques et projets urbains. Strates 13:1–11

Cazes G, Lanquar R, Raynouard Y (1995) L'aménagement touristique. Coll. Que sais-je ? n°1882. PUF, Paris

Colin C (2012, Septembre) La conquête du fleuve face aux sites industrialo-portuaires. Communication présentée aux 7e Rencontres de Mâcon, Fleuves et Territoires, Institut de recherche Val de Saône - Mâconnais, Mâcon

Di Ciommo F (2006) Métropolisation et planification urbaine dans une perspective de développement durable (Post-doctorat, CSTB le futur en construction). http://www.cstb.fr/fileadmin/documents/publicationsscientifiques/doc00004324.pdf. Accessed 2 July 2015

Drain M (1998) Régulation de l'eau en milieu méditerranéen: risques et tensions. Territoires en mutation 3:262

Giraud C (1994) Concepts d'une sociologie de l'action: introduction raisonnée. L'Harmattan, Paris

Gourlot N, Grissot C, Pain J, Puthod S (2009) L'eau en ville: approche environnementale, ouverture territoriale et gestion durable. In: Conference panel « Défis des villes durables ». Universidade Federal de, Pernambuco

Graillot D, Waaub J-P (2006) Aide à la décision pour l'aménagement du territoire: méthodes et outils. Lavoisier, Hermès Sciences, Paris

Gralepois M, Orillard F, Verdelli L (2018) L'instrumentalisation de la prévention des inondations dans les opérations d'aménagement, une justification de la gentrification ? Le cas de la ville-port du Havre (France), Carnets de Géographes, ongoing

Gravari-Barbas M (2004) La conquête d'une nouvelle frontière: réinvestissement symbolique et requalification fonctionnelle des fronts de fleuve urbains. ESO Travaux et documents 22:31–39

Grumillier F (2007) L'indécision: un modèle négatif pour l'action? http://www.philoflo.fr/resources/L$27ind$C3$A9cision.pdf. Accessed 2 July 2015

Humbert N (2014) Darsena di Milano – Projet spatial et influences d'acteurs urbains. Master thesis, University of Tours, Tours

Jacquet L (2013) Les valeurs dans le projet d'urbanisme. La figure du « tripode-culbuto » pour lire les valeurs du projet « Bossut » à Cergy-Pontoise. URBIA, Hors-série (1):37–52

Lechner G (2006) Le fleuve dans la ville. La valorisation des berges en milieu urbain. Note de synthèse pour la DGUHC, Direction générale de l'Urbanisme, de l'habitat et de la construction

Marié M (1983) Pour une anthropologie des grands ouvrages. Le canal de Provence. Les Annales de la recherche urbaine 21(1):5–35

Marié M, Tamisier C (1982) Un territoire sans nom: pour des approches des société locales. Librairie des Méridiens, Paris

Paffoni E (2012) Historique de la relation ville-port en Ile-de-France. Rapport du Programme de recherche FLUIDE, Agence Nationale de la Recherche, Université Paris Est, Unité Systèmes Productifs, Logistique, Organisation des Transports, Travail

Pierron J-P (2006) L'urgence. In: Guineret DH, Carvallo S (eds) L'action: délibérer, décider, accomplir. Ellipses, Paris

Rode S, Verdelli L, Gralepois M, Bonnefond M, Orillard F (2015) Concevoir et fabriquer un quartier en zone inondable: comment les projets urbains intègrent-ils le risque inondation? Cas du Havre, Angers et Narbonne Conference panel « Concevoir et fabriquer la ville », Rencontres internationales en urbanisme – APERAU, Rennes, 1–5 June 2015

Romain F (2010) Le fleuve, portuaire d'images urbaines: formes et enjeux. Géocarrefour 85(3):253–260

Salamon J (2008) « Pédagogie de la ville », Les mots des acteurs dans le projet urbain. Certu, Coll. Débats, Lyon

Verdelli L (2008) Héritages fluviaux, des patrimoines en devenir – Processus d'identification, protection et valorisation des paysages culturels en France, Portugal et Italie: quelques exemples significatifs. PhD Thesis, University of Tours and University of Coimbra

Verdelli L (2014) Port heritage as a material support for port-city's identity. In: Dans water and cities, managing a vital relationship, conference proceedings of the 50th ISOCARP congress "urban transformations, cities and water", Gdynia, pp 1424–1434

Verdelli L, Morucci F (2014) Renouveler l'identité de la ville entre culture portuaire et loisirs. Le cas de Livourne Loisir et Société/ Society and Leisure 37:58–78

Vermeersch L (1998) La ville américaine et ses paysages portuaires entre fonction et symbole. L'Harmattan, Coll. Géographie et cultures, Paris

Zittoun P (2013) La fabrique politique des politiques publiques: une approche pragmatique de l'action publique. Presses de la Fondation Nationale des Sciences Politiques, Paris

Grenoble, the River City Facing the Mountains (End Nineteenth Century-1930s)

Michaël Attali and Natalia Bazoge

Summary

At the end of the nineteenth century, the function of rivers changed and they were integrated into the urban environment through the practice of physical, recreational and leisure activities, which gave them a new meaning. This redefinition contributed to transforming the representation of the river "from a hostile element to an area of discovery and relaxation" (Levet-Labry, VertigO – la revue électronique en sciences de l'environnement, Hors-série 10, 2011). Representations and river activities developed in a specific way in the city of Grenoble as a result of the presence of a further spatial and structuring natural element, the mountain environment of the Belledonne, Chartreuse and Vercors massifs. Analysis of the representations shaped by Grenoble's river, the Isère, appears relevant in defining the latter's place and function within its territory. While the Isère established itself as a pole of attractiveness, its topography nonetheless created obstacles for the organization of festive and sporting events. Consequently, the leisure activities set up contributed as much to enabling its visibility, as to highlighting the difficulties in gathering crowds of people there.

Keywords

Isère · Grenoble · Rowing · Swimming · Territory

M. Attali (✉)
University Rennes, Laboratoire VIPS2, Avenue Charles Tillon, CS, Rennes, France
e-mail: michael.attali@univ-rennes2.fr

N. Bazoge
Université de Grenoble Alpes, Laboratoire SENS, Grenoble, France
e-mail: natalia.bazoge@univ-grenoble-alpes.fr

Introduction

At the end of the nineteenth century, when rivers as trade routes entered into competition with the railway, their function within urban space was gradually redefined and they became part of both leisure practices, such as canoeing as a form of entertainment specific to the bourgeois society (Vivier 1994), and tourist exploration activities (Hajek 2007). This situation was of particular interest in the case of the Isère. From the 17th and 18th centuries onwards, the economic role of the "half river, half torrent" Isère was not comparable to that of the Rhône or the major waterways in the west of the kingdom. Navigating the river nevertheless occupied an important place in the provincial economy until the end of the eighteenth century, despite particularly difficult navigational conditions (Dubourgeat 1989). Locally, navigation enabled the transportation of a range of goods between the small ports along the river, and at provincial level, it constituted a link between the main sites of the Dauphiné's metallurgical industry. In the nineteenth century, however, natural hazards, as well as the development of rail links between the Grenoble and the Rhône Valley, substantially reduced the commercial advantage of the river, which now had only a negligible economic role within the region (Favier 2010). At the same time, the increased urbanization of Grenoble encouraged the development of new areas conducive to strolling. The Ile Verte promenade along the banks of the Isère was therefore created in 1862, under the influence of the Parisian model of the Parc des Buttes-Chaumont and Bois de Vincennes (Baret-Bourgoin 2010). Flow control of the river through the construction of dams upstream from the Alpine city was essentially intended to protect Grenoble from flooding, but also allowed new ways of developing the Isère, in the same way as the construction work on the banks of the river undertaken in 1835 (Parent 1982). The setting-up of sports societies and the creation of various forms of entertainment in the last decade likewise attested to the appropriation of the Isère for leisure purposes. The representations and

C. Machemehl et al. (eds.), *Reclaiming and Rewilding River Cities for Outdoor Recreation*, Estuaries of the World,
https://doi.org/10.1007/978-3-030-48709-6_9

activities linked to the river nonetheless developed in a singular way in the city of Grenoble on account of a further spatial and structuring natural element, the mountain environment of the Belledonne, Chartreuse and Vercors massifs. In terms of tourism, for example this mountainous area was the focus of the various associations engaged in promoting the region and they gave priority to its development and accessibility, even though the city of Grenoble was considered to be an essential detour and embodied the Alpine city or "capital of the Alps" (Baret-Bourgoin 2010). Furthermore, the plan to develop and extend the city put forward by Léon Jaussely and adopted by the city council in 1925 did not include work to valorize the river, apart from the construction of bridges intended to facilitate contact with neighbouring towns and villages. On the contrary, increased emphasis was placed on the mountain area with the creation of the Parc de la Bastille on the lower slopes of the Chartreuse massif at the edge of the city (Blanchard 1925), followed by the inauguration of the cable car in 1934. In light of this topography, the question of the Isère's structuring role in the development of urban space may be pondered (Favier 1993), in the same way as the Rhône which, for its part, was perceived as a constraint (Delahaye 2004).

Analysis of the representations constructed in relation to this river thus appears to be relevant in defining its place and function within the territory. In flowing through the city against a background of Alpine peaks, did it take its place in the continuity of the mountains' representations and symbolize the purity of its source, thus enabling the development of a hygiene-related role (Vigarello 1985)? Or, on the contrary, in crossing an industrialized valley (Dalmasso 2008) and urban territory, did it contribute to diverting some away from the river? The evolving representation of the river with regard to the surrounding mountain territories resounded with the image of the city of Grenoble, particularly from a touristic point of view. The study of representations makes it possible to situate the perception of an area that was long perceived as a natural hazard (Le Lay and Rivière-Honegger 2009) despite being tamed. Consequently, while the Isère established itself as a magnet, it nonetheless suffered from a degree of mistrust shown by a population, which needed time to apprehend it. As a result, the leisure activities established led as much to making it visible as to highlight the dangers of crowds of individuals gathering there. Over and beyond these representations, this paper also considers the resistance encountered in the construction process of a river city in terms of recreational sports.

This study is based on the analysis of the regional daily press (*Le Petit Dauphinois*) and that of the local sports press (*Les Alpes Sportives*), in order to define the representations. It likewise rests on the analysis of municipal archives so as to situate the integration of the Isère in the projects and realizations undertaken within a sporting and municipal framework.

Structuring Space

As from 1856 onwards, boatman Pierre Martin was appointed lifeguard in charge of supervising activities near the river between 1 May and 30 September. This appointment constituted a key moment in the possibility of developing the Isère. It highlighted the interest of the municipal authorities for a place that could henceforth be adopted and which should be given particular attention. In fact, while aquatic space may procure certain pleasures, it presented at the time a major disadvantage in relation to other environments, namely the high risk of drowning given the very low number of people who knew how to swim. In the case of the Isère, this risk was accentuated due to the strength of the current, the various obstacles and the debris carried along by the river (Dubourgeat 1989). It was therefore the water's environment more than the water itself which would be the focus of the first efforts undertaken to develop the river banks.

The end of the nineteenth century was indeed marked by a strong desire to convert the banks of the Isère, whose three-stage evolution may likewise be observed in other spaces (Collins et al. 2008). While the river had hitherto first been considered as dangerous, then secondly as insalubrious, the third stage corresponded to the taming of an area mainly composed of wasteland so as to turn it into an outdoor living space that was fully fledged and integrated into the city. The organization of merchant markets was also an opportunity to reflect upon the commercial areas now taken over by leisurely strolling along the Isère. During discussions on 5 June 1886 concerning the organization of one such market, the city council studied the best site allowing exhibitors to set up their stands in an area mainly unused but with strong potential for a period of 6 years. Far from being temporary, these discussions displayed the will to commence an urban evolution based on a river that had become a structuring element of the city which had hitherto tended to turn its back on it.

The first indicators of the global development of the river that was to be an integral part of urban management appeared at the end of the nineteenth century. The creation of the Isère lifeguard society, the *Société des sauveteurs de l'Isère*, on 9 July 1883 was an important indication of the new stakes taking shape. While it was in line with the recruitment of Pierre Martin for the rescue of persons in distress, this society also envisaged engaging in both the provision of training programmes to attract new groups of people to water activities and the development of events aimed at making attractive an environment still on the fringes of urban dynamics. These initiatives underlined the wish to secure an accident-conducive area at a time when political will to develop it was emerging. Equipped with boats financed by the city of Grenoble, this society had the task of ensuring safety on the riverbanks. It would also participate in bringing life to the environment. The night festival, *La fête de la nuit de l'Isère*, held on 4 October 1896, had ambitious aims. It aspired to bring families together in a natural setting so they would better know the banks, as yet not fully developed. As an honorary president, Grenoble's mayor showed how the Isère was perceived as a constituent pole of attractiveness for the city of Grenoble. In the same vein, the national federation of swimming and lifesaving societies, the *Fédération nationale des sociétés de natation et de sauvetage*, requested the help of the Mayor of Grenoble in organizing its national festival. The programme included swimming and lifesaving contests, interschool swimming championships, an Army and French Navy swimming competition and a fully clothed Army and French Navy lifesaving championship. The Mayor's speech on 10 December 1909 underlined the importance of the event in the recognition of Grenoble as a major centre of tourism and sport, which now included not only its peaks, but also the promotion of its aquatic assets. Finally, the proposal put forward by Mr. Gigarel to the Mayor concerning the recognition of Grenoble as a health resort following the city council's session of 25 January 1912 showed a desire to turn the city towards its natural assets. The rapporteur highlighted the gentle climate and its privileged location, which should contribute to making it a stronghold of tourism. The quality of life implied the valorisation of this reality which should be based on the river flowing through the city. The Isère and the Drac were mentioned as two key geographical features that largely explained its *dolce vita*.

Yet the river also had a structuring role for space around Grenoble. While the city was the central magnet through the festive events produced, the river created an element of urban dynamics that went beyond the limits of the city: "The trumpeters of Grenoble, led by their amiable president, Mr. Drevot, took part in this water expedition and travelled in the first two boats, followed by the guests and the honorary and active members of the lifeguard society. After a most pleasant journey along the enchanting banks of the Isère, leaving behind successively la Porte de France, Pique-Pierre, Sassenage and Orlandière, the squadron arrived (. . .) within sight of the pont de Veurey bridge; the trumpeters of Grenoble then played one of the best pieces of their repertoire. A large number of curious spectators came to watch the picturesque landing of our marines who, forming a line and humming the music in their heads, marched proudly into the village under the command of Mr. Clarenson (. . .)" (Municipal Archives of Grenoble 1903). Flowing not only through the city, but also more especially a territory, the river participated in connecting cities and contributed to creating a social link based on shared recreational affinities. It played a role in opening up the nearby area to populations that were still very much enclosed within restricted and confined spaces.

Entertainment

While the river was instrumental in structuring urban space, the will to appropriate the Isère also resulted in a multiplication of forms of entertainment intended for the inhabitants of Grenoble. The end of the nineteenth century was characterized by the proliferation of a variety of newly created societies and events organized around the Isère. The *Union nautique de Grenoble*, Grenoble's water sports association, was therefore set up on 29 December 1885 with the aim of "developing the physical and moral strength of its members through the practice of swimming and canoeing". Founded on 21 September 1893, *L'Aviron Grenoblois*, Grenoble's rowing society, equally attested to the attractiveness of the Isère for the organization of new forms of practice and sociability. Supported by the greatly increasing modernity of sport at the end of the nineteenth century, the players in the process took over a new territory, an El Dorado that was to be conquered and transformed into an arena for the expression of new forms of sociability. The water sports festivals organized by this society in 1909, 1911 and 1914 formed part of this perspective. Alongside the sporting event, festive entertainment was also programmed, such as concerts and pigeon releases, in order to "generate enthusiasm among the people of Grenoble for all aspects of watersports" (*Aviron Grenoblois* 1911). The purpose of the *Aviron Grenoblois* was explicit in its aim to open up and appropriate a new territory: "We have taught the young people, members of our society, employees, French and foreign students, about the beauty of our sport, we have revealed to them the many charming secret spots of the Isère upstream from Grenoble. In a word, we have contributed to creating greater appreciation of the soil of our little homeland by showing the inhabitants, even those living in Grenoble for a long time, a new dimension of the beauty of our dear Dauphiné" (Annual Report of *Aviron Grenoblois*, 1907–1908, p. 3). This society played an important role in the development of activities related to the river, which was intended to become a place where life happened. And that was how the multiplication of swimming and lifesaving contests made it possible to gather crowds in hitherto previously abandoned space. Acclaimed in 1901 when awarded a prize by the President of the Republic, these events attested to the growing influence of the river on the forms of sociability developing in and around Grenoble.

During the 1920s, this momentum continued. The *Aviron grenoblois* had its own boathouse on the banks of the Isère and used the river for its training sessions. Various contests were held between the Grenoble teams,[1] although the aim displayed as from 1920 was to set up regional regattas on the Isère. They were held for the first time in September 1920 (*Les Alpes Sportives*, 18 September 1920), then again in October 1921 (*Les Alpes Sportives*, 15 October 1921). They enabled the grouping of teams from different regional clubs in a competitive event. In the same vein, the local swimming club, *club de natation des "Dauphins"*, created in 1920, also used the river for its water festivals, in particular the Paul Bisch Challenge,[2] alongside the regattas of the *Aviron Grenoblois*. The challenge was as much to develop health and hygiene objectives through the sport as to initiate a festive attraction surrounding the Isère and the water sports resort it constituted. Thus, a comment was invariably made during the organization of the *Aviron grenoblois* regattas concerning the reception of spectators, positioning of the stands and areas of lawn on the banks.[3] Pleasure and good humour were always associated with the representations conveyed concerning the activities organized around or in the river. From the "merry" water sports festival of the Isère lifeguards' swimming school to the "happy" rowing competitions, symbolic positivity took shape concerning the waterway which consequently acquired unprecedented status. During the regattas of the *Aviron Grenoblois*, the "amusing nautical interludes" prepared by the lifeguards of the Isère, for example made it possible to sustain the festive atmosphere. This appropriation approach towards the river was reinforced by the gradual implementation of nautical tourism on the Isère following the creation of an Alps branch of the *Canoë-Club de France* in Grenoble. This branch organized cruises on the Isère, particularly on the stretch between Albertville and Grenoble (*Les Alpes Sportives*, 15 April 1922), and in *Alpes Sportives* published a "canoeist's guide to the Isère", which detailed the various steps and potential difficulties in navigating the river (*Les Alpes Sportives*, 2 September 1922).

Furthermore, the local press saw these various events as a way of making the inhabitants of Grenoble aware of a variety of sporting activities. The swim across Grenoble traditionally held on 14 July, this "great sporting event of nautical propaganda" (*Les Alpes Sportives*, 1 July 1922), for example enabled spectators to discover not only swimming, but also other activities such as diving and water polo. It very quickly became an organized demonstration of the range of activities offered by the club. Swimmers going down the river livened up their journey with a number of water polo passes and the best divers were invited to perform acrobatic feats from the wooden suspension bridge over the Isère (*Le Petit Dauphinois*, 15 July 1924). In this way, appropriation of the river played its part in the sporting acculturation approach implemented in the local and regional press at the beginning of the interwar period (Bazoge et al. 2012). In 1925, the daily newspaper *Le Petit Dauphinois* underlined both the presence of a "considerable crowd" and the size of the event, "since the taste for water and rowing seems suddenly to have swept over a number of Grenoble's young inhabitants" (*Le Petit Dauphinois*, 5 August 1925).

The navigability conditions of the Isère nonetheless raised recurrent issues concerning the management and appropriation of this space. For both swimming and rowing, the Isère lost some of its attractiveness as a result of its difficult navigability conditions. For swimmers, the main problems concerned the temperature of the water and the current. As from 1922, the difficulty in organizing a swimming race on the Isère was highlighted: "The result would have no meaning whatsoever on account of the unevenness of the swimming lanes, with the Isère flowing much more quickly along its convex bank. The only race possible in the Isère was a timed race with staggered starts, each swimmer being compelled to follow the river axis; it stands to reason however that such a race would be of little interest to the public" (*Les Alpes Sportives*, 8 July 1922). While demanding conditions may also represent sources of interest for spectators as a result of the uncertainty surrounding swimmers' progress "in the cold and rapid waters of the Isère, full of various and often unexpected difficulties" (*Les Alpes Sportives*, 1 July 1922), they also required a high level of vigilance to avoid accidents. The *Société des Sauveteurs de l'Isère*, therefore ensured the swimmers' safety, yet doing so required appropriate know-how and equipment. In the case of rowing, the current likewise proved to be a discriminating factor in competitions. As a result of these practically unmanageable conditions, the Grenoble regattas failed to reach the same level as other regional competitions. And so, the competitive stakes which appeared increasingly significant within sporting societies during the interwar period conditioned a deviation vis-à-vis the river. Although the population was henceforth more familiar with water activities, it was necessary to redirect events towards the quest for sporting performances dependent on normalized conditions. Consequently, the *club de natation des "Dauphins"* henceforth organized competitions in a private pool managed by a firm from Grenoble or at the public swimming pool (Fig. 1). As a result of the swimming conditions in the private pool, the club was nominated to organize the swimming championships of the South East in 1921 (*Les Alpes Sportives*, 9 July 1921). A

[1] For example in 1920, three teams of the *Aviron grenoblois* competed in the Challenge of the *Ligue Maritime Française*, French Maritime League. *Les Alpes Sportives* dated 15 May 1920.

[2] The Challenge included a 200-m downhill breaststroke event, a cross-Isère event and a 400-m downhill freestyle event. *Les Alpes Sportives* dated 16 July 1921.

[3] For example in 1921, covered seating with a capacity of 300 and two lawn areas were set up on both banks at the finishing line. *Les Alpes Sportives* dated 8 October 1921.

Fig. 1 Exposition internationale, Grenoble 1925

new pool was made available to the club in 1925 "equipped with a shallow area" allowing the frolics of a great number of beginners" (*Les Alpes Sportives*, 9 May 1925). As for *Aviron grenoblois*, the Isère became a site that was less conducive to the organization of its events as a result of competition from Lyon's navigable rowing areas and the nearby lac du Bourget lake. The club's regattas were moved to the lac de Laffrey, contributing to the valorisation of a new territory. In fact, they were added to the programme of another sporting event, the Laffrey hill motor race (*Le Petit Dauphinois*, 15 July 1926). Changing the location allowed greater sporting mobilization around these regattas with the presence of the largest regional clubs (*Le Petit Dauphinois*, 17 August 1925). This change may also be considered as the reflection of a focus that was being increasingly directed towards the slopes, thus tending to reduce the river's attractiveness which had been reactivated since the mid-nineteenth century. It was not before the 1970s and the emergence of new forms of recreational sport that renewed interest towards it could be seen.

Conclusion

The initiatives undertaken in the recreational taming of the Isère rested on a redefinition of the representations linked to the urban waterway. They contributed to changing it from a hostile space to one of discovery and relaxation without, however, overlooking the risks associated with an environment that still posed social problems because of limited swimming instruction and was therefore difficult to develop. This was all the more true in the case of difficult natural conditions, as with the Isère. By bringing people together within the framework of various events, these initiatives participated in the appropriation of the fluvial space and resulted in the Grenoble territory no longer being exclusively structured around its peaks. For this reason, it is possible to consider that the pioneers of sport from Grenoble led to the landscaping (Niel and Sirost 2008) of the Isère. The work conducted here attests to the potentially structuring role of the river within the urban project. This analysis shows how rivers played a part in urban reorganization as from the nineteenth century and inaugurated a new type of public space (Romain 2010). Nevertheless, the increasingly prominent competitive stakes throughout the 1920s led sporting societies to gradually turn away from this river that was hard to control because of its difficult navigation conditions, in favour of facilities that were more adapted to performance. This called for other ways of appropriating the space, in order for it to continue to exist in face of the mountain peaks. It was not until new recreational forms of expression were socially recognized that the Isère once again became a place for leisure activity. Moreover, during the interwar period, the river regained its central economic role as a hydroelectric resource, not only at local level, but also as part of a national network (Dalmasso 2008). The "Hydroelectric Power and Tourism" poster for the World Fair, organized in Grenoble in May 1925, attested to this repositioning of the representation and use of the waterway, in the face of tourism that was once again essentially focused on the mountains.

References

Archives municipales de la ville de Grenoble. Cote 3R45, Sociétés sportives nautiques et de sauvetage, 4 octobre 1903

Aviron Grenoblois. Courrier à Mr le Maire de Grenoble, 02 mai 1911

Baret-Bourgoin E (2010) De nouveaux espaces, une nouvelle identité urbaine. In: Favier R (ed) Grenoble. Histoire d'une ville, Glénat, pp 120–128

Bazoge N, Attali M, Grosset Y, Delorme N (2012) The regional press as an instrument for sports acculturation: a case study of Les Alpes Sportives in the interwar years. Int J History Sport 29(8):1159–1176

Blanchard R (1925) Le plan d'extension de Grenoble. Revue de Géographie Alpine 13(3):667–670

Collins T, Muller E, Tarr J (2008) Pittsburgh's three Rivers. From industrial infrastructure to environmental asset. In: Mauch C, Zeller T (eds) Rivers in history. Perspectives on Waterways in Europe and North Americ. University of Pittsburgh Press, Pittsburgh

Dalmasso A (2008) Barrages et développement dans les Alpes françaises de l'entre-deux-guerres. Revue de Géographie Alpine 96 (1):45–54

Delahaye E (2004) La dialectique des villes et du Rhône à l'aval de Lyon: des villes malgré le fleuve? Urbanisation et contrainte fluviale. Géocarrefour 79(1):85–93

Dubourgeat J-P (1989) Une rivière et des hommes : aperçus sur les gens de l'Isère aux XVIIIe et XIXe siècles. In: La ville et le fleuve, 112e Congrès national des Sociétés Savantes, Lyon, 21–25 avril 1987. CTHS, Paris, pp 253–272

Favier R (1993) L'Isère et son rôle dans l'économie et l'organisation de l'espace régional du XVIIe au XIXe siècle. In: Piquet F (ed) Le fleuve et ses métamorphoses, Colloque international, Université Lyon 3, 13–15 mai 1992. Paris, Didier Erudition, pp 183–193

Favier R (2010) Franchir ou naviguer sur l'Isère. La rivière dans l'espace régional (XVIIe-XIXe siècle). La Pierre et L'Ecrit 21:115–133

Hajek S (2007) Histoire culturelle d'une société nautique: le Canoë kayak club de France 1904–2004. Thèse de l'Université Paris Descartes

Le Lay Y-F, Rivière-Honneger A (2009) Expliquer l'inondation: la presse quotidienne régionale dans les Alpes et leur piedmont (1882–2005). Géocarrefour 84(4):259–270

Levet-Labry E (2011) Aménager les rivières et réduire les risques pour développer le tourisme nautique en France (1904–1924). VertigO – la revue électronique en sciences de l'environnement, Hors-série 10

Niel A, Sirost O (2008) Pratiques sportives et mises en paysage (Alpes, calanques marseillaises). Etudes rurales 2008(1):181–202

Parent J-F (1982) Grenoble, deux siècles d'urbanisation. Grenoble, PUG

Romain F (2010) Le fleuve, porteur d'images urbaines: formes et enjeux. Géocarrefour 85(3):253–260

Vigarello G (1985) Le propre et le sale : l'hygiène du corps depuis le Moyen Âge. Le Seuil, Paris

Vivier C (1994) L'aventure canotière du canotage à l'aviron. Histoire de la Nautique bisontine (1865–1930). Thèse de l'Université Claude Bernard Lyon 1

The Role and Significance of the Recreational Reconquest of Port Spaces: Rouen (France) Reinvention at the Neck of the Estuary

Damien Féménias, Olivier Sirost, and Barbara Evrard

Summary

Rouen offers a unique opportunity to study the evolution of large-scale port wastelands in a metropole located at the neck of an estuary, the most landlocked of the great European seaports. What is the role and significance of recreational conversion in public policies? The chapter puts the significance of this conversion and the role of these practices into perspective by situating them in the context of the central, network and nodal relations between the city and the port. A new urbanity centred on the port is being established, revealing different usages, publics and regions that give expression to physical and symbolic discontinuities that are reflected in original narratives.

Keywords

Port-city · Sports leisure · Estuary · Conversion · Recovery · Wastelands · Media narrative

Introduction

For a variety of different reasons, the great ports of the Western world (Vérot 1993) are reconfiguring their installations, restructuring and repositioning their activities and generating considerable areas of abandoned wasteland in metropoles that are often already saturated. When these spaces become available, they provide opportunities to refashion cities both politically and in terms of urban planning.

The city–port relationship is already the subject of many studies highlighting the economic importance and the role of the different spaces comprising them (e.g. Boyer and Vigarié 1982; Brocard et al. 1995; Ducruet 2008; Hoyle 1989; Pultrone 2008). However, these studies tend to ignore both riparian and upper estuary cities and the question of the physical and symbolic future of these abandoned port zones. These unique urban configurations are structured by three decisive factors: the central importance of the port in the city, the network structure it imposes on the city, and the nodality of this important crossroads (Ducruet 2008).

Given the range of possibilities, what is the role and significance of the recreational "reconquest" in public policies? What demands and expectations does their development respond to? What forms does it take? What practices does it involve? This contribution will answer these questions, based on the study of the most landlocked of the great European seaports, Rouen (France), a city that has undergone considerable transformations as a result of the downstream relocation of its port activities. The analysis is a considered overview based on a series of surveys conducted between 2007 and 2011 (Féménias et al. 2011), financed by the downstream Seine public interest group (GIPSA) in an attempt to comprehend the different experiences of the estuary produced by leisure activities in order to develop an original and informed appreciation of its social functions.[1]

The chapter situates these developments (that extend the city centre) in their context of interregional development, and goes on to recount leisure usages in the concrete configuration of Rouen, which filters these activities and imposes to and fro movements on those it tolerates. Finally, we shall retrace the political and media narrative that gives a central role to the river and constructs a plot in an attempt to give

D. Féménias · O. Sirost (✉) · B. Evrard
CETAPS EA 3832, Université de Rouen, Normandie Université, Mont Saint Aignan Cedex, France
e-mail: damien.femenias@univ-rouen.fr; Olivier.sirost@univ-rouen.fr; barbara.evrard@univ-rouen.fr

[1] Four data collection campaigns conducted over 5 years, with reference to two principal and complementary scientific programmes (Usages récréatifs et Sequana): http://seine-aval.crihan.fr.

C. Machemehl et al. (eds.), *Reclaiming and Rewilding River Cities for Outdoor Recreation*, Estuaries of the World,
https://doi.org/10.1007/978-3-030-48709-6_10

meaning and coherence to a region that is in the throes of transformation.

At the Centre Lies the Port and, at the end of the Estuary... the State

The etymology of the name Rouen, *Rotomagos,* "the market on the river", testifies to the central and foundational role of the port in the metropole. It is currently one of the 3 trading ports[2] in the Seine Axis, giving shape to and developing the "Grand Paris" project of extending the Île de France region to the sea by means of the Seine and all the way to Le Havre (Attali 2010). Initially inseparable, the city and the port gained in strength and autonomy in the course of the twentieth century (1966), becoming physically separate. The dynamic of the port led it to extend and relocate downriver (11 terminals distributed over 100 km). The port finally relocated out of the city in 1960.

Given the size of the freed-up areas (150 hectares, several km of quays in the heart of the city), the possibilities for development were enormous and have been played out all along the Seine estuary since the early 1960s (Anquetil 2002), under the aegis of the State and its instrumental authority, the EPBS (a public land management body). The aim was to prepare local communities to "receive Paris without being subjected to it", by encouraging them to develop a culture of prospective real estate development by equipping them with recycling capacities capable of absorbing wastelands in order to "reconstruct a city within the city".

The transformations of Rouen thus form part of the regional development programme, described as a "reconquest", designed to promote the rational reorganization of estuary zones. To this end, the eco-districts in city centres and the Open Air and Leisure Bases (BPALs) on urban fringes offer residents consistent leisure structures that are both complementary and strategic, developing the city by "gardening nature" without impacting on the economic corridor (Féménias et al. 2013). What is happening in Rouen (west city centre), in Caen (on the peninsula) and in Le Havre (southern districts) forms part of one and the same interregional programme that seeks to reconcile these cities with their maritime environment (Anquetil 2002).

The case of Rouen presents one original characteristic (Decoux et al. 2010) in that it has no sizeable heritage[3]: no vast docks-warehouses, unlike London, Antwerp or Le Havre and no remarkable urban façade, unlike Bordeaux (Decoux et al. 2010: 105). This absence offers an opportunity to extend and rethink the city centre. Its needs have been identified and, since 1972, the choices have been made: they consist of constructing offices and housing and re-injecting life into the centre city by means of a major economic turnaround towards trade, tourism, leisure and culture[4] (Guermond 2008). As presented, the eco-districts form part of a more global urban narrative of "recreational reconquest" of the port area. This "regeneration" (described variously as "redynamising", "making the city within the city", "embracing the logistic heritage" and "forgetting petrochemicals") combines the economic turnaround of the city as it attempts to reconcile these different aspects in infrastructures that give symbolic expression to a lifestyle that is redesigned and restructured with local services (Renauld 2012): public transport (buses), shops (the Docks 76 shopping mall), the upper service sector (state-of-the-art office buildings), educational (H20, panorama XXL) and sporting facilities (Kindarena, small private gyms in hangars) and socio-cultural activities (106 and hangar 23).

A Bowl-shaped Configuration, Structured for and by Flows

Moreover, Rouen functions as a multimodal platform (Berque 2000). This network configuration, which implicates the port city in multiple networks that are always difficult to interconnect, imposing a hierarchy of flows. In addition, the meandering configuration of the valley and the bowl-shaped morphology of the Rouen area make terrestrial traffic almost impossible by establishing Rouen as an essential crossroads that finds all road and rail traffic converging on the centre city.

With a majestic and omnipresent river, the setting consists of rolling hillsides that limit and define the horizon, and a badly adapted road network that situates major and very busy transportation routes between the city and the lower quays. Begun in 1998, the recreational conversion of the lower quays of the Seine is still incomplete: whether linear or "on the surface", they provide an adapted physical reality and constitute a dramatic context for wandering cyclists, joggers, skaters, etc. (Escaffre 2005).

---

[2] https://www.haropaports.com.

[3] The choices made for the post-war reconstruction of the city eliminated a remarkable urban facade along the banks of the Seine, a "sister to the Bordeaux facade, now classified on the UNESCO world heritage list", raised the level of the quays, distributed over two levels, and devoted them to automobile traffic and rail freight.

[4] Rouen is a pioneer for protection and conversion of its heritage. As early as 1964, it used the Malraux law to create a protected sector; in 1970 the rue du Gros Horloge became a pedestrian zone, and in 1972 Rouen became a "City of Art and History". This know-how currently finds expression in new forms, with cultural tourism and festivals, most notably *Normandie Impressionniste*.

The larger areas host the most impressive events: the funfair (Foire Saint Romain[5]), the tall ships festival (Armadas[6]), summer concerts,[7] the seaside decor and "beach" events designed for the summer holiday period (Rouen sur mer[8]) form part of a cultural offer that mobilizes considerable means (advertising, trade and security). For the moment the quays themselves only allow for linear activities and are presented as developed promenades. The vast majority of the population, frequenting the riverbanks consists of strollers, walkers, runners, cyclists or roller skaters. The city is structured by and for logistics, right down to its recreational development, with these systems generating to and fro movements and flows. Like the port, the rehabilitation and conversion dynamic for the quays are towards linear extension: beyond Rouen, the long-term plan is to line the whole river with green lanes. The stretches left fallow on the edges of the city are still often not accessible for cycling and do not yet provide circuits for those who wish to wander outside the city (Féménias et al. 2012).

Water-based leisure activities are very restricted by law in Rouen, being intensely localized, contained and monitored (Féménias et al. 2011). With freighters and barges coming and going, there is no room for sailing, which involves tacking. The river bed is very narrow and changing currents make sailing on the river a very technical exercise for all. In 1935, the trading port uprooted recreational activities: although it agreed to the opening of a winter harbour in 2008, it still ruled out the organization of regattas for "Sunday sailors" in the city centre. Water sports in the Rouen area are thus rectilinear with to and fro movements and organized in river zones, on Lacroix Island with rowing, canoeing and kayaking clubs and public bathing. Although these practices build up very different perceptions of the environment, overall, the river is still greatly appreciated by all and the risk of collision (with barges) constitutes, along with the fear of a ban on these activities, a commonly shared preoccupation. In May, the 24-hour powerboat race, an endurance race for teams, attracts nearly 400,000 spectators to the river banks of the riverside city. But a fatal accident in 2009 highlighted the limits of a voluntary organization which, in spite of its great popularity and repeated success, has not convinced the authorities of the importance of suspending all commercial traffic during the event. Based on misappropriation of port equipment, water sports are only tolerated.

A Maritime Character Showcased to Recount the City and Create a Region

Autonomous but inseparable, the city and the port of Rouen intertwine interests that are difficult to reconcile. At the hub of town planning questions, the Seine is also the central character around which the local political narrative is constructed: the conversion of the river banks involves all the players in the region in a new functional and symbolic contract. The narratives concerning the Seine Axis, Grand Paris (Greater Paris) and the reconquest of the river constitute a veritable ongoing media series. The operation also sees itself as fostering resilience, proposing to dress a symbolic wound by bringing the city closer to the river, which had grown more distant in the wake of post-war reconstruction. The river provides local representatives and journalists with a story to tell and a way of recounting both the region and its inhabitants.

The removal of the port also offers politicians an opportunity to leave their own mark on the town planning and architecture of the city centre. Liberated from all port heritage that might be considered cumbersome,[9] the field of expression is nevertheless saturated with references to the port that a post-industrial aesthetic imposes on the whole: cranes and hangars take on an "ornamental" function (Barthes 1957), contribute to a "décor effect" that lends an original echo to current music ("Le 106"), contemporary dance (Hangar 23), science (hangar H20) and the fine arts (panorama XXL, in the shape of a silo). The customs warehouse and the power plant converted into the "docks 76" shopping mall retain the names,[10] dimensions and materials (concrete, brick, glass and steel) of the original creations. The architectural (post-industrial) and town planning (speaking of "eco-districts" rather than "buildings", for both up-market housing and "state-of-the-art" offices) choices give a global coherence and tangible form to the overall "narrative" which the politicians make a point of promoting in order to underscore the significance of the developmental evolution of the metropole.

The port narrative and post-production decor entirely devoted to consumption—whether cultural or commercial—provide something for everyone, offering each a way of living and recounting the space corresponding to their own experience. Rouen *sur mer* (on the sea) enables the local authorities to reinterpret for themselves the images of the

[5] This month-long fun fair has been held yearly in the city for nearly 900 ans.

[6] A gathering of tall sailing ships that is held every 5 years.

[7] These concerts can attract as many as 70,000 people (04 July 2014).

[8] A local version of *Paris Plage*, it consists in bringing the seaside (sand and decor) and seaside events to the heart of the city during the summer vacation period.

[9] Only the tide gauges have been protected as historic monuments since 1997: for nearly a century they have embellished the hydraulic system with tall towers that empower the cranes with their lifting capacity and magnify the institutional power of the Chamber of Commerce with the original architectural features they provide for maintenance installations.

[10] Cast in concrete, the indication "*entrepôt des douanes*" (customs warehouse) can still be seen on the main building of a shopping centre known as the "docks".

equitable and media policy of *Paris Plages*. *Paris Plages* (or Paris Beach) is a summer operation that Paris town hall has conducted since 2002. During the school holiday period a part of the riverside expressways is closed to automobile traffic. The riversides are landscaped (e.g. with sandy beaches, lawns, palm trees, deckchairs, showers and sprays) organized and come to life (open-air cafés and restaurants, artistic events and sporting facilities) and most of all showcased as a rediscovery by the local population and a social redistribution of seaside pleasures to those who do not go away on vacation, as well as a gesture against pollution and the reign of the automobile. The recreational "rediscovery" of the river acts as vector for a communication campaign whose principal mission is to reduce the role of the automobile in the city. The great media success of this Paris operation has made it a model that is very largely adopted and imitated in both France and abroad. The legendary economy of this policy and political narrative (the event is orchestrated and staged by professionals), like the *Armada* festival (Féménias et al. 2013) and the BPAL (Evrard et al. 2013), brings the sea (an imaginary sea, a representation of the sea, an ornamental sea) to those who have no access to it, and enables local politicians to promote the city in the media. The "business eco-districts" (Luciline), the *Normandie Impressionniste* festival and major infrastructures (e.g. Flaubert Bridge, Kindarena, Le 106, H20) provide the Metropole with vectors on its own scale that enable it to exist differently[11]—making it a new institutional and political player, and promoting both the quality of life and its representation. The department makes the river the guiding spirit for a more global offer (BPAL, coast), that weaves summertime solidarity (free ferries, "read on the beach" operations, "the coast for €1"). The Chamber of Commerce and Industry, the Metropole and the Region use these décor elements to promote the other economic industrial and port-based realities of the Seine Axis. The estuary narrative works because it is effective, easy to produce, to mediatize-broadcast,[12] grasp and integrate in order to give meaning to a changing world. The maritime aspect is a way of dealing with these changes and differences, which are constructed by the port logic on a daily basis, and offer a means of imagining a felicitous globalisation and deindustrialisation.

Conclusion

Rouen represents a unique opportunity for all who wish to understand the contemporary redefinitions of city–river relations, which are so many markers of our condition. The Seine flows through and fashions a system of cities, forms part of changing usages and contributes to defining evolving lifestyles. For almost 20 years, the recreational conversion of the quays has stood as a milestone and been hailed as an indisputable success: the quays have become part of everyday life (Di Méo 1996), significant milestones for the local representatives who promote them and for the populations that frequent them. The functional and symbolic contract on which new living spaces and new lifestyles are established illustrates and showcases a local narrative whereby the city's history as a port acts as a symbolic balm that assuages fears and veils the brutality of deindustrialisation and the attendant risks,[13] thus offering a vision of the future and establishing a lasting bond between the populations and their environment.[14]

This approach combines symbolic desires for a change of air and scenery: turning back to the river testifies to adjustments that support the reintroduction of nature into the city (Augustin 1995), but in terms of proximity rather than a continuation of the city centre. These leisure activities open the way for localities to act as mediators of social relations (Di Méo 1998), and confirm the previously observed and growing demand for natural spaces in the field of French sporting activities.[15]

Apart from land-based activities, water-based activities are very much contained and only barely tolerated. Different attitudes to these usages produce the potentially great cultural, social, territorial and ecological discontinuities of future transformations. We may well wonder whether these practical and territorial disparities which generate margins and marginalities do not contain the seeds of future conflicts or proposals calling for new systems of enhancement and new forms of governance (Féménias et al. 2013).

References

Anquetil P (2002) L'Établissement Public de la Basse Seine. PCM Le Pont 1:43–48

[11] Previously the Metropole managed water, transport and waste services. Now, with these installations it leaves its stamp on the city and enhances its image with the festival.

[12] Many books, magazines, journalistic programmes and various embellishments have been produced, demonstrating the many commercial or individual conversions possible through practices and objects.

[13] Coming to bear on a regional economy that is still largely constructed on the transportation of petroleum and its industrial transformation.

[14] Pollution of the river, limiting its bed, quality of water, reduction of humid zones and biodiversity.

[15] Eurobaromètre survey, 2014, by the European Commission: http://europa.eu/rapid/press-release_MEMO-14-207_fr.htm.

Attali J (dir.) (2010) Paris et la mer. La Seine est capitale. Fayard, Paris

Augustin J-P (1995) Sport, géographie, aménagement. Nathan, Paris

Barthes R (1957) Mythologies. Le Seuil, Paris

Berque A (2000) Médiance, de milieux en paysages. Belin, Paris

Boyer JC, Vigarié A (1982) Les ports et l'organisation urbaine et régionale. Bulletin de l'Association des Géographes Français 487:159–182

Brocard M, Lecoquierre B, Mallet P (1995) Le chorotype de l'estuaire européen. Mappemonde 39:6–7

Decoux J, Couchaux D, Gaillard G, Kollmann C, Miossec Y (2010) Le port de Rouen. Région Haute Normandie, Inventaire général du patrimoine culturel, Rouen

Di Méo G (dir.) (1996) Les territoires du quotidien. L'Harmattan, Paris

Di Méo G (1998) Géographie sociale et territoire. Nathan, Paris

Ducruet C (2008) Typologie mondiale des relations ville-port. Cybergeo (online)

Escaffre F (2005) Les lectures sportives de la ville: formes urbaines et pratiques ludo-sportives. Espaces et Sociétés 122:137–156

Evrard B, Féménias D, Sirost O (2013) Les bases de plein air et de loisir en Haute Normandie, un ersatz de bord de mer? Etudes Normandes 1:81–88

Féménias D, Sirost O, Evrard B (2011) Les loisirs nautiques dans l'estuaire de la Seine. Médiations territoriales, consciences du milieu. VertigO, special edition no 10

Féménias D, Evrard B, Sirost O (2012) Les loisirs sportifs dans l'estuaire de la Seine: natures périurbaines et cadre de vie. Loisir et sociétés 34(1):67–98

Féménias D, Evrard B, Sirost O (2013) Une Armada de signes. Mise en scène et publics des grands voiliers. *Etudes Normandes* 1:31–48

Guermond Y (dir.) (2008) Rouen, la métropole oubliée? L'Harmattan, Paris

Hoyle BS (1989) The port-city interface: trends, problems, and examples. Geoforum 20:429–435

Pultrone G (2008) Trieste: new challenges and opportunities in the relational dynamics between City and port. Méditerranée 111:129–134

Renauld V (2012) Fabrication et usage des écoquartiers français: éléments d'analyse à partir des quartiers De Bonne (Grenoble), Ginko (Bordeaux) et Bottière-Chénaie (Nantes). Unpublished PhD thesis, INSA de Lyon

Vérot P (1993) De la crise des ports au renouveau des villes littorales. Mappemonde 1:40–43

Recreational Activities, Economic and Territorial Development: Caen (France) in the Reconquest of its River?

Sébastien Bourdin and Yann Rivoallan

Summary

Like numerous port cities in France and in the world that undertook a reterritorialisation of their industrial–port wastelands, Caen is seizing the opportunity of such a move. Its development should be carried out within the framework of an integrated territorial approach, through a quality urban and landscaped treatment and an urban animation where leisure combines cultural and sporting events. Since the project is still in its early stages, this chapter proposes a geoprospective trial on the development of Caen-to-the-Sea Peninsula through recreational and sporting activities. It starts, in particular, with the recently adopted Master Plan and reviews, in light of what has been done in other cities around the world that have embarked on operations to reconvert and revitalize their port areas by using them as new tools of economic development (Das, Regional Outlook 92–95, 2011), and thus, makes some recommendations on the future of this project.

Keywords

Territorial development · Waterfront · Urban revitalization · Leisure activities

Introduction

Following the introductory work of Jansen-Verbeke (1986), the waterfront model can be considered as an element of urban tourism which also helps to reduce the climate and energy footprint (Lecroart and Palisse 2007) by imagining new solutions, as part of a sustainable approach (Lecroart 2007). In particular, conceived as a mean of developing leisure tourism, it can play an important role in the attractiveness of a city (Lecroart et al. 2007). The waterfront has been widely discussed in the literature and there is considerable evidence that, when it is redeveloped for entertainment and recreational purposes, it helps to improve the image of the city and is a real economic development tool (Butuner 2006; Soraya 2009; Das 2011). In this case, the regeneration of brownfields has an important role to play in the economic development of these areas, attracting tourists and locals, national and international residents with high purchasing power (Griffin and Hayllar 2006; Craggs and Schofield 2011). It is designed to serve as a catalyst for the quality of life in the city (Savage et al. 2004; Erkok 2009) by acting as a social cohesion tool and participates in the emergence of new urban experiences in connection with the cultural and territorial heritage of the sites concerned (Brutomesso 1993; Hoyle 2001; Hamzah 2002; Giovinazzi 2010). It allows the inhabitants of the territory to reappropriate areas (in particular, banks and quays) that had been confiscated from them for decades or even centuries (Lecroart and Palisse 2007). However, it is unfortunate to note that the French urban planners have lagged far behind in this area (Prelorenzo 2010).

It is within this science-based context that we wish to examine these issues, so that a city like Caen can reconnect with its river through recreational and leisure activities by proposing a geoprospective trail (Gourmelon et al. 2012) around a river: the Orne River.

Caen has a millennium past in maritime and commercial history. The first Industrial Revolution, combined with the opening of the Caen Canal to the sea at the origin of a peninsula (see Fig. 1) and the arrival of the railway, provide to the port of Caen, as well as its region, a significant economic growth. Like many European ports, the maritime, port, industrial and commercial activities have steadily declined as

S. Bourdin (✉)
EM Normandie Business School, Métis Lab, Caen Cedex, France
e-mail: sbourdin@em-normandie.fr

Y. Rivoallan
Caen Normandy Metropolitan District, Caen, France
e-mail: yrivoallan@em-normandie.fr

C. Machemehl et al. (eds.), *Reclaiming and Rewilding River Cities for Outdoor Recreation*, Estuaries of the World,
https://doi.org/10.1007/978-3-030-48709-6_11

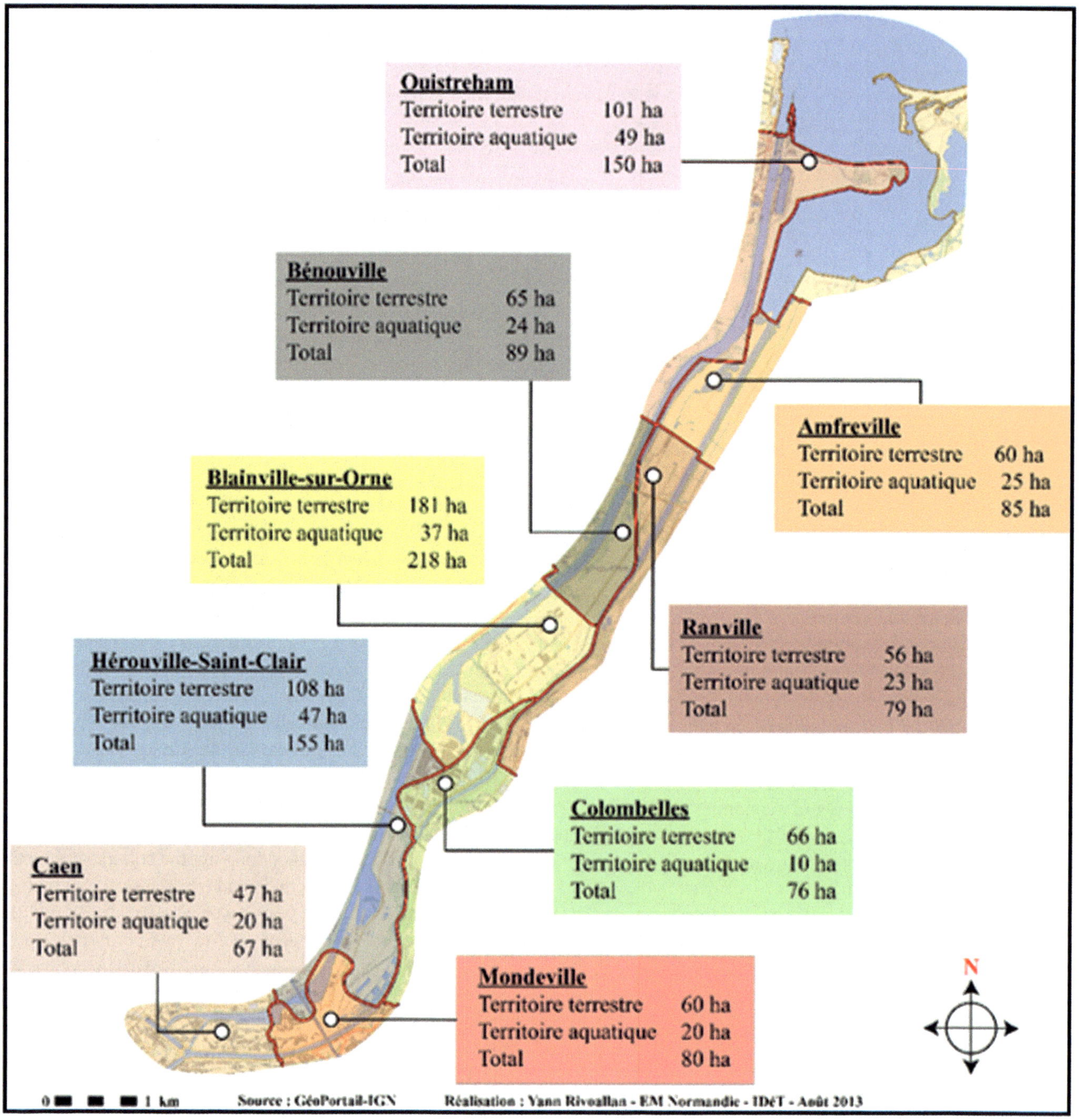

Fig. 1 Development potential of Caen-to-the-Sea Peninsula's municipalities. Source: Géoportail, IGN—réalisation Y. Rivoallan

a result of the containerisation of goods and the recurrence of economic crises since the late 1970s, leading to the creation of a large wasteland, from Caen to Ouistreham and going through nine municipalities.

The fragmentation of the urban environment, resulting from the desertion of industrial activities, durably transforms the landscape, on the one hand, but also the urbanity of these cities on the other hand, resulting in the representations the inhabitants have, their ownership and their images. As a result, this is a substantial challenge for municipalities to carry out urban revitalisations. The first industrial-port wastelands revitalisations date from the 1960s in the United States and aimed to develop declining urban territories, which became very high added value land reserves (Halbert 2007), freed up from their former port activities relocated to more efficient sites. In line with this, many French cities and in the world have carried out an upgrading of their industrial-port wastelands (Lechner 2006; Lucchini 2012; Rivoallan 2013). For instance, REVIT (revitalisation of industrial wastelands) European programme is co-financing European cities projects with the objective of rehabilitating industrial wastelands in a sustainable manner. Revitalisation concerns a former freight station in Stuttgart, industrial and port sites in Nantes and Medway, coal mines in Torfaen, a former textile factory in Tilburg and the Hart van Zuid in Hengelo. In Nantes, for example REVIT has contributed to reviving three former sites (Alstom, Atlantic foundries and former shipyards) on the island of Nantes, facing the historic heart of the city, inspired by the memory of the places and valuing the industrial heritage. The long-term expected result is the construction of 6500 housing units, 250,000 m^2 of economic activities, open public spaces, social, tourism and cultural facilities.

Caen did not fully take the opportunity of such an approach even though its inner Peninsula represents a unique source of economic development for the territory. Nevertheless, this area, whose vocation is to be re-territorialised and re-converted, cannot only limit itself to being a new retail catchment area. Its development must be carried out within the framework of a sustainable development process, through quality urbanism and landscaping and urban animation where leisure activities combine cultural and sporting events, whilst preventing this area from becoming only a gentrification place (Chaline 1988; Raoulx 1996). However, for the time being, only a small part of the Peninsula has been developed and everything else in this huge area between Caen and Ouistreham will have to be reviewed.

First, our proposal will be to assess barriers and opportunities that led to the partial development of this land reserve meant for economic and territorial development purposes. Second, we will not only suggest geoprospective ways (Gourmelon et al. 2012) of development to this large area, but will also make some recommendations for its future development.

Reconquer the Orne River Through the Reconversion of Wastelands: Principles and Limits

The Maritime Past: The Valuable Asset of a River Town ...

The maritime, industrial and commercial past of a port must be integrated into a process of urban requalification in order to preserve the identity and the memory of places, varying according to the social groups involved, within an approach that is at the same time economic, cultural and environmental (Chang and Huang 2005). Taking into account the maritime heritage of these abandoned industrial port areas is part of the development strategies implemented by the local authorities (Raoulx 1996). All stakeholders must play a role in the decision process on what should be preserved or not, and in the ways the places will be reappropriated (Chang and Huang 2005). This is especially necessary if one wants to (1) facilitate cohesion among the different social groups, (2) link the past with the present and the future and (3) facilitate the collective reappropriation of these places. However, the latter is only possible if heritage is not sanctified or imported (Gourbin 2012).

From this point of view, Caen has undeniable assets related to its history that local populations could recapture through a project they would be proud of (Chang and Huang 2005). Unlike its Norman sister cities of Rouen and Le Havre, Caen has never had an important port activity. Being more of a commercial town than a town with a maritime tradition (Raoulx 1996), Caen has always been relying on an intense short sea shipping activity since the beginning of the middle Age. It is during the fifteenth century that the port actually increased activity throughout Europe with the development of trade in fine draperies and Caen stone for construction. Its successful activity will lead the city to carry out a succession of developments (Carel 1886) eventually resulting in the creation in 1857 of a peninsula with a port vocation close to the City Centre (Raoulx 1996).[1] The opening of the Caen Canal to the sea as well as the inauguration of the railway station provided a significant growth to the city and its port which then became both a land and a maritime crossroad (Loraux 1946; Pinson 1985) at the beginning of the twentieth century.

It was at this time that the Port of Caen considered moving from a commercial function to an industrial one. It was in 1903 that the Société Navale Caennaise shipping line was created. As soon as the First World War was over, the industrialisation of the port actually began with the creation of the military shipyards of Blainville-sur-Orne and the opening of the Société Métallurgique de Normandie (SMN) steel mill in Colombelles (Raoulx 1999). Until the end of the 1980s, the Lower-Normandy region that is very rich in iron[2] ore ensured the Port of Caen, an important return freight to the ships coming from the United Kingdom, Belgium, the Netherlands and Germany loaded with the coal needed for the SMN steel works activity.

However, with the shutdown of the SMN in 1993, it is more than 50% of the port's activity that is de facto

[1] In the meantime, the port will experience ups and downs, particularly with the introduction of the Fontainebleau Edict (1685), revoking the Edict of Nantes (1598), which will have a devastating impact on Caen. It will ruin the majority of local businesses owned by Protestants who represented at this time between 30 and 50% of the Caen population. It will be followed by the dissolution of local shipping lines created during the previous century, stopping the development of the port.

[2] Domestic second largest producer after the Lorraine region.

Table 1 Size of the area to be revitalised within the Caen region

Municipality	Occupied area within the Caen-to-the-Sea Peninsula			Ratio
	Land area (ha)	Water area (ha)	Total (ha)	% of total surface area
Caen	47	20	67	6.7
Mondeville[a]	60	20	80	8.0
Hérouville-Saint-Clair[a]	108	47	155	15.5
Blainville-Sur-Orne	181	37	218	21.8
Colombelles	66	10	76	7.6
Bénouville	65	24	89	8.9
Ranville	56	23	79	7.9
Amfreville	60	25	85	8.5
Ouistreham	101	49	150	15.1
Total	744	255	999	100

Caen-to-the-Sea Peninsula's municipalities-Territories-Superfaces-Y. Rivoallan, August 2013

[a]Not including the "Nouveau Monde" sewage plant area

disappearing. Under pressure from regional agricultural lobbyists and despite the close proximity of the Port of Rouen, the largest European port specialised in cereals traffic, the Caen port authorities decided to invest in the construction of grain silos to develop grain traffic (Raoulx 1996) as an alternative to the predicted decline in port activity. Nevertheless, the volume of this activity is low, insignificant and steadily declining and the continuing decrease in activity of the Port of Caen for several decades has turned it into an industrial-port wasteland with considerable reconversion potential for recreational purposes.

As we have seen, the area identified from Caen to the sea went through different stages following the model developed by Lucchini (2012) of territorialisation (industrialisation), deterritorialisation (abandonment or relocation of industrial and/or port activities), and reterritorialisation (revitalisation of the industrial-port brownfields). Nevertheless, this last process seems to encounter in Caen several difficulties that must be analysed.

. . . But Barriers We Should Not Ignore

In 1857, the inauguration of the canal of Caen to the sea created an area of about 15 km long commonly referred to as the Peninsula-of-Caen-to-the-Sea (Fig. 1). It starts from the Saint-Pierre Basin in the Caen Harbour and leads into the Bay of Sallenelles, between Ouistreham, Sallenelles and Merville-Franceville. This strip of land, an inner peninsula, is bordered on its left by the canal of Caen to the sea and on its right, by the Orne River, a small coastal river having its source near Sées in the Orne Department and terminating in the English Channel. This continuous territory (Fig. 1), over nearly 15 km, passes through many municipalities: Caen, Hérouville-Saint-Clair, Mondeville, Colombelles, Blainville-sur-Orne, Bénouville, Amfreville, Ranville and Ouistreham. This area of 1000 hectares is both an opportunity and a challenge for the agglomeration of Caen and the lower Normandy region, insofar as the potential for development of a wasteland of this size is exceptional in Europe. For example, the Glasgow-Clyde waterfront represents 660 hectares and the Amsterdam-IJburg is 400 hectares, yet these revitalisation projects have been very successful (Tables 1 and 2). Nonetheless, for Caen, behind this potential for development, real difficulties arise connected with industrial risks and the governance of the project.

The Caen section is the smallest surface of this area and its development is stalled by industrial hazards within the territories of the peninsula belonging to the municipalities of Mondeville and Hérouville-Saint-Clair. The Technological Risk Prevention Plan (PPRT) prohibits, in particular, any urban development within a several hundred-metre perimeter exclusion zone.

Industrial hazard exposure is particularly prevalent within the upstream part of the Peninsula (Caen, Mondeville and Hérouville-Saint-Clair) due to the presence of hazardous companies (Dépôts Pétroliers Côtiers, Les Combustibles de Normandie, SOFRINO, AGRIAL) and also because of the constant movement of hazardous materials[3] via road, rail and waterway. Hence, a PPRT (SEVESO II high-risk classification) regarding Dépôts Pétroliers Côtiers was approved in 2015. All these industrial risks act as a "lock" preventing any revitalization with territorial continuity of this large area. Caen is, thus, obliged to develop its part of the peninsula outside its geographical limits, Hérouville-Saint-Clair is forced to transform its area of the peninsula into a high-density truck traffic zone and to keep a large untouched wasteland and Mondeville forced to limit its urbanisation operations.

[3]Gasoline and Class 3 flammable liquids, Class 1 explosives, Class 7 radioactive materials, fertilizers, gases and pesticides (PLU-Overview report-Arrêt-28 January 2013, p. 234).

Table 2 Main urban regeneration areas in Europe

Current or planned wasteland regeneration projects	Area in ha
Amsterdam-Ijburg	400
Barcelone-Forum Besos	216
Bilbao-Zorrozaurre	60
Copenhague-Ovestad	310
Glasgow-Clyde Waterfront	660
Londres Stratford City	1450
Rijeka-Delta Barros	62
Rotterdam-Stadshaven	1500
Swansea-SA1	51

Current or Planned Wasteland Regeneration Projects—Y. Rivoallan, August 2013

Yassin et al. (2013) investigated the governance role in the development of waterfronts in Malaysia. The results showed that low involvement and limited collaboration among the stakeholders were contributing factors of project inefficiency hindering successful completion. Yet another barrier connected to the governance of this project is that while the stakeholders seem to be convinced of the existence of Caen's maritime, industrial and commercial past and that some heritage buildings and equipment must be protected, the multiplicity of stakeholders involved does not foster on the consistency of a global revitalisation project. While in most European projects, the number of municipalities is very limited, nine municipalities (Fig. 1) are part of the Caen-to-the-Sea Peninsula and directly concerned by the industrial-port brownfield. As a result, it leads to governance difficulties in this joint project. Hence, the Société Publique Locale d'Aménagement de la Presqu'île-de-Caen-à-la-mer (SPLA), a local public company for the development of the Peninsula of Caen-to-the-sea, was created to carry out the task of drawing up the development project of the Industrial-Port Peninsula but did not associate all the municipalities from Caen to the sea. Only Caen,[4] Hérouville-Saint-Clair and Mondeville, the agglomeration of Caen-La-mer, the Basse-Normandie region and Ports Normands Associés (PNA) are involved in this revitalization through the SPLA, excluding the six closest municipalities to the sea. In addition to this, divergent views exist among the mayors of Caen, Hérouville-Saint-Clair and Mondeville concerning the process of regeneration of this area. The result is a fragmentation of the Peninsula into separate functional areas corresponding to sub-projects of development over a period of time and space without a global project or "umbrella-project" emerging and reaching consensus because of lack of genuine consultation and not taking into consideration the overall needs and wishes of the Peninsula's nine municipalities.

[4] The city of Caen has a major position.

The River Orne as a Differentiating Advantage: Géoprospective Elements

Feedback on the River as a Differentiating Advantage

The river can be considered as a territorial asset in the sense of Pecqueur (2005),[5] i.e. it represents a potential for the territory to be revealed or to be organised as it does not exist as such. Also, the cities that regenerated their waterfront have created a real differentiating advantage (Pecqueur 2006, 2007) owing to this asset which is the river, allowing them to stand out in the competition among other cities. In other words, the territories concerned used water and industrial-port wastelands nearby to create an economic and territorial development[6] resource by attracting high purchasing power tourists and local, domestic and international residents (Griffin and Hayllar 2006; Craggs and Schofield 2011). Reappropriation of brownfields for recreational purposes also meets a need for nature and a growing quest for well-being in the city (Bourdeau-Lepage and Roland 2013). The river and the river ecosystems are, from this point of view, a privileged territorial and ecological resource and the revitalization of these areas is an asset for the territories in competition with other cities.

The reappropriation of this territorial and ecological resource needs substantial financing, involves many stakeholders (state, regions, municipalities, investors, business leaders, local residents, tourists, boat owners and associations) and must benefit from a visible strategy, a flexible, thoughtful, detailed, analytical and continually reassessed Master Plan or Guide Plan to avoid urban, social, ecological and economic mistakes that can be catastrophic (Lecroart et al. 2007; Giovinazzi 2010). In particular, all future waterfront social and ecological functions and its day-time and night-time activities as well, must be considered at the earliest stages of the initial analysis of the project (Giovinazzi 2010). In other words, building a real differentiating advantage is quite often a complex process but this urban singularisation has become essential in terms of territorial attractiveness. It should not only be seen as the essential to attract new inhabitants and businesses but also to keep them in the territory. By revitalising a waterfront, the city proposes a new territorial offer that meets the expectations of the various economic agents, making it more attractive.

[5] Researcher at the laboratory "Territories", UMR 5194 PACTE, UJF/UPMF/CNRS, Professor at the University Joseph Fourier, Deputy Director of the Institut de Géographie Alpine.

[6] Through the creation of new high added value residential, commercial, recreational, nautical and cultural uses.

A waterfront development should enable regenerating and reinventing of wastelands, with creativity, dynamism and modernity, with the aim of designing an effective economic development tool for the region (Raoulx 1996), keeping local residents within the territory, attracting domestic and international tourists and creating new jobs in various areas (Craggs and Schofield 2011). The territorial project should be able to provide residential housing based on social diversity, unspoilt environment, leisure, shopping facilities, local services, tertiary activities, marinas and bring together people from the territory around a project they can be proud to call their own (Chang and Huang 2005).[7] The success of such projects is based on the respect of the river–land–sea continuum.

The Peninsula's Development Project

Envisioned in the time of Jean-Marie Girault, Mayor of Caen from 1970 to 2001, the development of the Peninsula-of-Caen-to-the-sea has been taking shape for some years now. The almost finalised Local Urban Planning Plan project, dated 28 January 2013,[8] plans a significant urban densification of the town with the construction of 1400 dwellings per year until 2016, half of which located downtown. It expects the peninsula to become an exemplary residential area, in terms of urban planning and environment, by recovering in a sustainable manner banks and water, focusing on yachting, leisure, shopping and promenades and also, reducing parking and motor vehicle traffic by favouring carbon-free vehicles. In addition, the city of Caen made a commitment in May 2010 to curb its emissions of CO2 by 20% by 2020 and to take into account the vulnerability of the Peninsula, related to road transport of dangerous materials and also, that of industrial risks.

Several public buildings were already built in this part of the Peninsula—the Cargö[9] and the Art & Media School. Two other buildings have just been built: the Regional Multimedia Library (BMVR) and the new Tribunal. With regard to environmental matters, a large 2000 m^2 lawn stretches out in front of the Court, up to the Victor Hugo Canal (Fig. 2), thus creating a continuum between the river, its banks and further downstream, the sea.

Beyond developments being finalised at the tip of the Peninsula, the MVRDV architectural firm extended the scope of the project by widening the area to be developed beyond the geographical boundaries of the Peninsula (Fig. 3). The subsequent Guide-Plan was called "The Great Mosaic" and is predominantly based on the respect of existing structures and provides a "realistic, attentive and user-friendly" urban planning. Proposed projects involve transforming the former factories into a collection of gardens to "aerate the area", a highly connected mobility plan mainly based on soft mobility, a number of new pedestrian walkways and a better interconnection with the sea. This latter ambition is to give meaning to the name of the Communauté d'Agglomération "Caen-la -Mer". The project will start near the historic centre of Caen and will follow the Orne valley that connects the city to the sea, not far from the D-Day beaches. The transformation has the ambition to radically improve the attractiveness of Greater Caen. The Guide-Plan will last more than a decade for its implementation but only concerns the municipalities of Caen, Hérouville-Saint-Clair and Mondeville, again without considering the other six communes alongside this long strip of land mainly due to political reasons. Yet, the inclusion of all the municipalities having a geographical connection with the area to be regenerated would allow new intercommunal cooperations, limit the risk of territorial breaks (Lechner 2006) and limit conflicts relating to land use. Furthermore, it should be noted that the Guide-Plan does not take into account the existing industrial risks (Seveso II high-risk classification for one of them) that are hindrances to the development of the Peninsula (Fig. 4) and the solution to relocate these high-risk industrial infrastructures is not addressed politically.

Upgrading the Peninsula of Caen-to-the-Sea: Between Managerial Analysis and Geoprospective Trial

Recent geoprospective approaches (Gourmelon et al. 2012; Houet and Gourmelon 2014) aim to better integrate, by using various methods, the territory into the process of exploration of the future, going beyond a simple illustrative support function of its possible futures. In this context, this approach requires a clear understanding of the whys and wherefores of the projects, in this case referring to the reterritorialization of a large industrial-port area and its integration into a sensitive space where rivers, estuaries and coastline areas are interrelated. As such, we are going to suggest ways that should be integrated in the planned development of "The Great Mosaic" project.

Urban Planning

The urbanisation of industrial-port wastelands offers many benefits. On one hand, it makes it possible to carry out significant profitable real estate transactions, through densifying the space, by focusing on tertiary support

[7] Among others, see examples in Nantes, Rouen, Lyon, Copenhagen, Salford, Portsmouth, Bilbao, Ipswich and Dundee.

[8] PLU-PADD-Caen-28 January 2013.

[9] Modern Music Concert Hall.

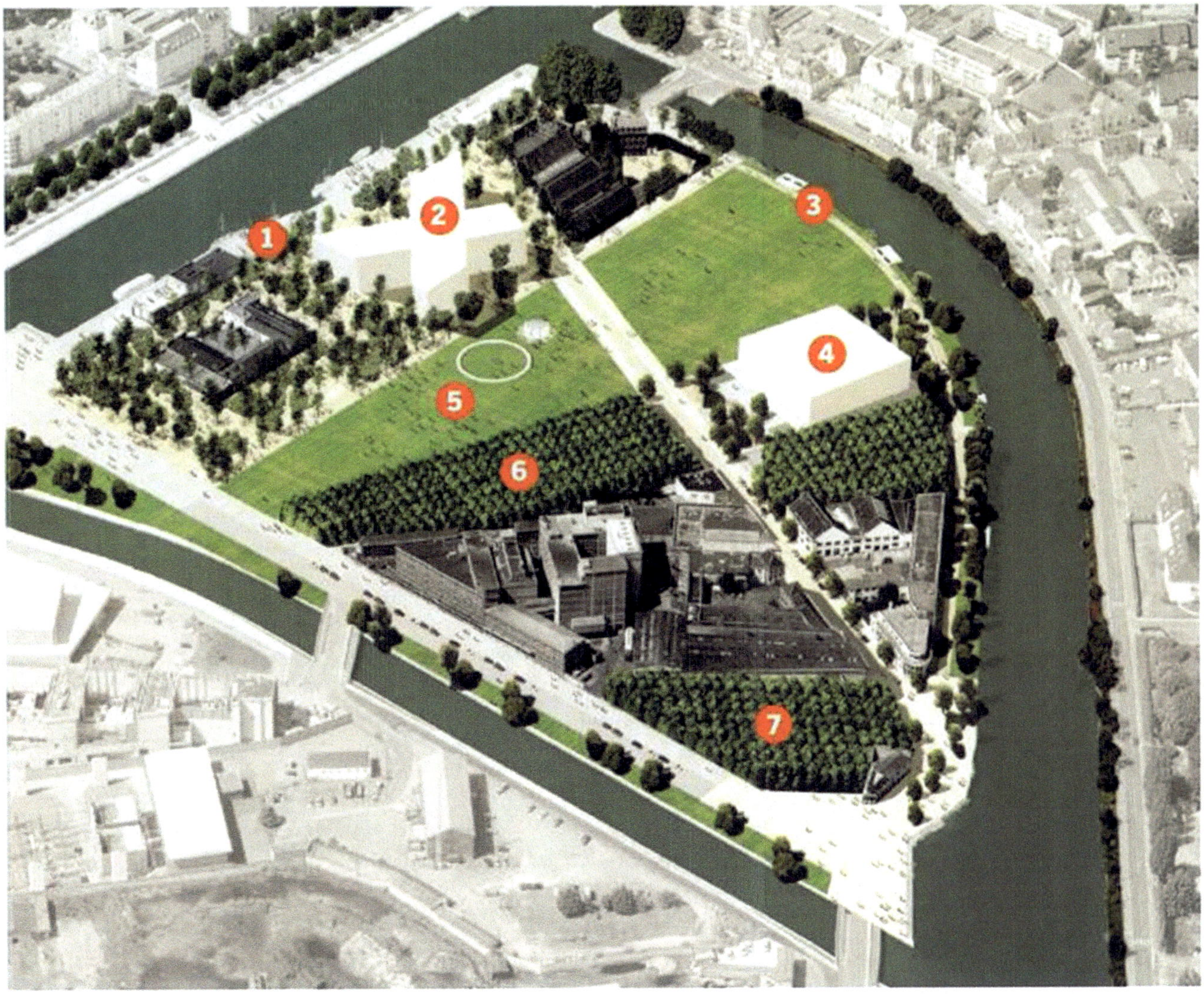

Fig. 2 Perspective view of the developments already carried out at the tip of the Peninsula. (1) Quai François Mitterrand, (2) BMVR, (3) Avenue Pierre-Berthelot, (4) Tribunal and (5) Lawn Michel Desvigne Group-Landscape architect. Source: MVRDV

activities, diversified urban functions[10] (Salet et al. 2007) and by housing various groups, in terms of age, family status, income and lifestyle (Tineke et al. 2007). On the other hand, these wastelands urbanisation help to fight urban sprawl. The agglomeration of Caen-la-mer can be considered "an expert" in urban sprawl in France even though it seems necessary to redensify the city. From this point of view, the Peninsula is an opportunity to bring people back to the city centre provided that the supply of housing is diverse both from the point of view of real estate price, size and housing quality. The objective is, therefore, to build a city district based on the memory of site and water so as to attract new residents and tourists. The developments will cover 200 of the 600 hectares of the Peninsula, resulting in 7000 dwellings built and 45,000 m^2 of economic activities created, in addition to the current existing situation. The urbanisation of these places should be able to develop vibrant neighbourhoods during the day and also in the evenings,[11] while avoiding an excessive gentrification of the area (Lecroart et al. 2007).

Zhang et al. (2011) studied recreational facilities' spatial structures in urban waterfronts. In particular, they showed how functional diversity and the creation of a "Green Belt" in parallel to the "Blue Belt" were the keys to the success of such projects. In the future project, consideration will need to

[10] Housing, leisure, yachting, university, services, offices and administrative functions in the city.

[11] A good example is GUNWHARFS QUAYS regeneration project in Portsmouth associating day and night life.

Fig. 3 Perspective view of future developments in the first part of the Peninsula. Source: MVRDV

be given to urban planning linking housing, community services, commercial activities and leisure.

From an urban planning point of view, the construction of an iconic and symbolic[12] building from the beginning of the project (Lucchini 2012) seems to act as cement; a flagship enabling the different social groups to reappropriate these places with new uses. We believe that it is imperative here to refer to the history and memory of these places. From this point of view, it seems relevant to us to consider the construction of such a building referring either to the history of Caen through the Second World War or to its industrial-port past.

Lastly, the consolidation or relocation of port operations allows the development of yachting by developing marinas within left vacant areas (Cremnitzer 2012) and allows a better interaction between water and land involved (Miller 2011). It is from this standpoint that a marina will be developed in Hérouville. Marinas generate jobs by bringing in visitors from outside the territory as well as families from the city, and by directly injecting money into the local economy. According to a 2007 study[13] about 15 direct and indirect jobs are created for 100 moorings and the marina draws an average of four visitors with overnight stays per boat. The marina is a must-see place to visit for 80% of families living in the city where the port is located. Finally, boat owners' expenses range from 175€ for a middle size boat and from 1800 € to more than 30,000 € (more than 24 m long yacht) for foreign visitors. These elements confirm the interest in developing this type of activity when it is associated with entertaining practices.

Economy

The redevelopment of an underused industrial port area, stemming from deindustrialisation and neglected for many decades, requires a complete rethink of its economic functions (Oakley and Rofe 2008) to have a positive impact

[12] Like the Spinnaker Tower in Portsmouth too with a sail shape chosen by Portsmouth people following an architects competition. It reflects the maritime history of the city known as one of the main military English port.

[13] Euromarina and Fédération Française des Ports de Plaisance (2007).

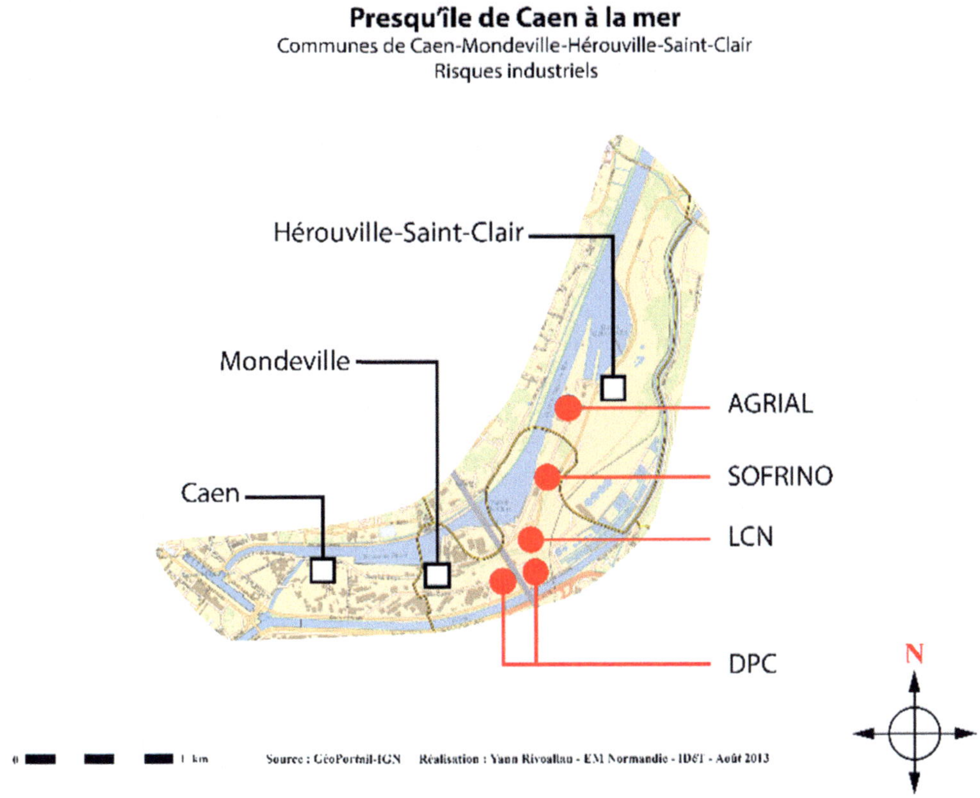

Fig. 4 Caen-to-the-Sea Peninsula-Municipalities of de Caen-Mondeville-Hérouville-Industrial Hazards-Y. Rivoallan-August 2013

on the territory's development. On one hand, physical transforming of degraded and polluted sites provides benefit to the local economy through important investments, create wealth and jobs owing to large construction projects. On the other hand, the successful regeneration of an industrial-port wasteland will attract companies and many tourists, which will benefit the local economy, generally in the retail, hotel industry, international trade shows and symposia, food and logistics industries sectors. Economic issues are, therefore, considerable in such a city as Caen when one knows the potential economic benefits of such projects.

The MVRDV architectural firm took into consideration the willingness of the SPLA partners to increase the number of cruises by improving the access and reception of cruise ships. This is a substantial challenge as competition with the English ports like Southampton and Portsmouth is real. The cruise sector continues to grow significantly with destinations to the Canary Islands or the Mediterranean from Southampton. This year, this city inaugurated its fifth cruise terminal and hopes to keep the monopoly of transatlantic cruises in Europe. But the proximity of Caen-Ouistreham to ports on the other side of the Channel is a real opportunity to draw cruise clientele, particularly, since a cruise passenger spends an average of 50 € locally before departure and such existing infrastructures are also an important opportunity to create jobs.

In order to achieve this economic development, supply must be in line with demand. It should be pointed out that favourite activities of tourists spending the day in waterfronts are related to shopping, restaurants (varied and of quality), bars and attractions/activities related to the water. For example, one can imagine developing kayak rides from Caen to the sea via the club, which is currently located close to the Peninsula, in the City Marina. Wandering on a large promenade alongside the banks and quays, the heritage sites, museums, galleries and all factors that bring in public must also be part of the action of consideration on future developments (Griffin and Hayllar 2006) and may become value-added sources for the territory.

Social

Beyond the purely economic aspect, Everard and Moggridge (2012) have clearly shown the acceleration of the reappropriation of the urban rivers in recent years in connection with the understanding of their landscape, ecosystem and social value. From this point of view, the revitalisation of an industrial-port wasteland acts as a social cohesion tool. It reunites territories separated by water canals, rivers and industrial and port areas (Giovinazzi and Moretti 2010). It modifies the urban landscape that has moved the city and its inhabitants away from these places that once represented employment and profitability (Chaline 1988). Furthermore,

Giblett and Samant (2012) have studied the sustainability of the successive development of urban regeneration programmes of waterfronts in some large Asian ports. In particular, they show the importance of giving greater consideration to social and environmental issues rather than only thinking from an economic point of view the development of waterfronts.

Facilitating the reappropriation of such areas deserted by the inhabitants is no simple matter. As such, memory of places plays an important role in the social approach of revitalised territories, but varies among the groups concerned and according to the experiences (Chang and Huang 2005). Here, we see two real issues for the municipalities involved in the development. The first is the future projectability to create social diversity. This, however, cannot be taken for granted, especially when it is a known fact that places associating heritage, tourism, leisure, quality living environment and boating aim more at the affluent social class than those in difficulty. The creation of new residential areas in the immediate proximity of a river with many landscapes, recreational and commercial facilities is all potential sources of gentrification. The second question is the ability of the project to foster interactions among the different stakeholders in these new areas (residents, tourists, developers, investors and workers). Indeed, the risk is important to see an addition of dense and diverse territorial practices but with limited exchanges among stakeholders.

One possible solution is to foster leisure activities that help reduce social inequalities in regenerated industrial-port Brownfields (Miller 2011). Yet, Caen-to-the-Sea Peninsula's significant size allows converting part of its area and its heritage buildings (silos) into places suitable for recreational activities: climbing, scuba diving, indoor skydiving, indoor surfing, water activities, theme park, sport and outdoor activities. These recreational activities could be social link catalysts.

Social dimension is first and foremost an urban planning issue, particularly related to housing typology. In other words, the aim will be to foster residential mix (large and small apartments, high-standard housing and social housing) and even intergenerational. When coming up with a project, this also requires consideration on the range of services offered (retail, tourism and leisure) and adapted to different levels of purchasing power.

Environment

The development of a revitalised area, generally speaking, of Caen-to-the-Sea Peninsula, in particular, is not possible without ensuring environment sustainability (Charlier 2007) and must contribute to the reduction of the climatic and energetic footprint (Lecroart and Palisse 2007), through the development of new solutions, within a sustainable-based approach (Lecroart 2007).

This approach requires decontamination and cleaning up of polluted soils (Lupi 2007), following the European Court of Auditors' recommendations based on the "polluter-pays" principle (Court of Auditors, EU, Report No. 23, 2012), improvement in the quality of water within revitalised areas (Giovinazzi and Moretti 2010), suppression or burial of high-voltage lines alongside banks and quays, development of sustainable neighbourhoods without cars (Lecroart 2007) or low-traffic urban areas (Lupi 2007), and considering environment preservation, from the initial period of the preliminary stage of the project. Finally, dealing with the problem of long-term rising sea levels must be our key concern.

Mobilities

The issue on mobility and difficult access to the Peninsula is very prevalent. There is currently no public transport service with the Peninsula and no direct connection to the sea. Moreover, due to very few "gateways", the connection of this area with its urban environment remains difficult. The reterritorialization of an industrial-port wasteland must be an opportunity to rethink mobility. The challenge is to reduce the carbon footprint and to allow the reappropriation of the site by the pedestrians, by planning from the early stage of the thinking process "carbon-free areas" without cars to reduce acoustic, olfactory and visual impacts. The project should, therefore, focus on green spaces and waterfront promenades, which are not only very popular with the residents and visitors, but also ensure continuity to the various places composing these spaces (Miller 2011). This nature restoration in the city is likely to guarantee ecological functioning restoration. Furthermore, the omnipresence of water in regeneration projects of areas located along canals or rivers is an opportunity to use the waterways to set up public transport that will participate in the comprehensive reorganisation of the urban fabric (Lechner 2006). Also, with regard to Caen, several issues concerning mobility must be rethought. The goal will be to imagine a shared mobility, to foster intermodal connections, to link North and South districts, to limit the Peninsula car traffic, to set up an environmentally friendly and alternative to the car, to optimise the existing infrastructures, to encourage the pedestrian traffic and above all, to connect Caen to the seafront.

The proximity of Caen and its Peninsula with the sea is an exceptional but insufficiently valued geographic advantage, especially from the perspective of sustainable development with an adaptive project planning to deal with climate change. The current urban grids network is generally focusing on cars. Pedestrian practice is difficult due to the preponderance of roads. It is, therefore, necessary to examine the modes of transport for both their efficiency and their optimisation. One could consider a Peninsula with a "quiet" bank, more associated with port imagery and related to the more residential and peaceful districts located on the River

Orne left bank hillsides and favouring soft traffic. It would result in the development of public spaces as landscaped promenades, cycle paths, playgrounds, gardens and parks close to the water and delimited by renovated and highlighted quays. The flat topography of the valley and the presence of a Green Lane towards Ouistreham are already existing advantages. Lastly, Caen and the other municipalities to the sea have an unused river transport potential. Following the example of other cities like London or Rotterdam, a possibility would be to set up a waterbus service linking Caen downtown to the sea and in particular, passing through Hérouville. The aim of this type of development would be to ensure a renewed relationship with the sea.

Conclusion

The municipalities of Caen-to-the-Sea are now facing the challenge of attractiveness. The revitalisation of such a large industrial-port wasteland such as the one being studied is an opportunity for the concerned territories. Nevertheless, major challenges remain, in particular, the one concerning financing and governance on the one hand, and the ecological perspective on the other hand. For the latter, indeed, building within an estuarine environment is a real challenge in the context of climate change and of gradual rise of sea level.

In recent years, the urban development and construction trend were to involve public sector and private financial resources to ensure feasibility and continuity of projects' implementation. Huang and Kao (2014) studied this matter in the context of regenerating waterfronts. In particular, they demonstrated how induced and balancing risk through the development of public–private partnerships depended on the participants' involvement in the project and on financial plans figured out to fund these types of projects. Today, it is generally accepted that public stakeholders must ensure from the very beginning, the quality of the project design as well as a genuine joint consultation with the community and the private stakeholders (Giovinazzi and Moretti 2010). These urban transformations require real cooperation taking into account the architects, investors, developers and the community's wishes (Desfor and Jørgensen 2004). Thus, for this project, the dialogue among the different stakeholders will have to be constant and open in order to create a climate of confidence encouraging initiative and risk taking. The winning strategy to reclaim areas coming back to life requires a worked out and sustainable territorial governance and a broad participatory initiative.

References

Bourdeau-Lepage L, Roland V (2013) Nature urbaine en débat: à quelle demande sociale répond la nature en ville? Déméter, 195–121

Brutomesso R (1993) Working on the water edge. In: Bruttomesso R (ed) Waterfronts a new frontier for cities on water. International Centre Cities on Water, Venice, pp 10–11

Butuner B (2006) Waterfront revitalization as a challenging urban issue in Istanbul. In: 42nd ISoCaRP congress: Waterfront Revitalisation as a Challenging Urban Issue, Spain

Carel P (1886) Histoire de la ville de Caen depuis Philippe Auguste jusqu'à Charles IX. Ed. Champion, Paris

Chaline C (1988) La reconversion des espaces fluvio-portuaires dans les grandes métropoles. Annales de géographie 97(544):695–715

Chang TC, Huang S (2005) Recreating place, replacing memory: creative destruction at the Singapore River. Asia Pac Viewp 46 (3):267–280

Charlier J (2007) Le caractère protéiforme des interfaces urbano-portuaires. In: Faire la ville avec le port, quelles stratégies pour le redéveloppement des espaces de liaison ville/port? Guide des bonnes pratiques, Paris

Craggs R, Schofield P (2011) The quays in salford: an analysis of visitor perceptions, satisfaction and behavioural intention. Int J Tourism Res 13(6):511–612

Cremnitzer J-B (2012) Réinventer l'existant: la créativité architecturale dans les friches culturelles. Édité par Université de Rouen-Atelier 231-La Friche-UMR CNRS IDEES

Das SB (2011) Singapore and the integrated resorts. Regional Outlook, 92–95

Desfor G, Jørgensen J (2004) Flexible urban governance. The case of Copenhagen's recent waterfront development. Eur Plan Stud 12 (4):479–496

Erkok F (2009) Waterfront: potentials for improving the quality of urban life. ITU A|Z 6(1):126–145

Everard M, Moggridge HL (2012) Rediscovering the value of urban rivers. Urban Ecosyst 15(2):293–314

Giblett G, Samant S (2012) A review of urban waterfront regeneration in global Asian port cities and the sustainability of their development. J Urban Regen Renew 5(3):266–279

Giovinazzi O (2010) Port cities and urban waterfront: transformations and opportunities, 3

Giovinazzi O, Moretti M (2010) Waterways and development of the territory: new mobility and sustainable tourism. Tema. J Land Use Mobil Environ 3(3)

Gourbin P (2012) Processus et actions patrimoniales dans les friches culturelles. Édité par Université de Rouen-Atelier 231-La Friche-UMR CNRS IDEES

Gourmelon F, Houet T, Voiron-Canicio C, Joliveau T (2012) La géoprospective, apport des approches spatiales à la prospective. L'espace géographique 41(2):97–98

Griffin T, Hayllar B (2006) Historic waterfronts as tourism precincts: an experiential perspective. Tour Hosp Res 7:3–16

Halbert L (2007) From sectors to functions: producer services, metropolisation and agglomeration forces in the Ile-de-France region. Belgeo. Revue belge de géographie 1:73–94

Hamzah A (2002) The concept of urban tourism, Kota Kinibalu City hall Kota Kinibalu, and Sabah Malaysia: paper presented at the seminar of the roles of government, Agencies on Urban Tourism, 29–30 January

Houet T, Gourmelon F (2014) La géoprospective–Apport de la dimension spatiale aux démarches prospectives. European Journal of Geography, Cybergeo

Hoyle B (2001) Lamu: waterfront revitalization in an east African Port-City. Cities 18(5):297–313

Huang W, Kao S (2014) Public–private partnerships during waterfront development process: the example of the world exposition. Ocean Coast Manage 92:28

Jansen-Verbeke M (1986) Inner-city tourism: resources, tourists and promoters. Ann Tour Res 13(1):79–100

Lechner G (2006) Le fleuve dans la ville. La valorisation des berges en milieu urbain. Centre de documentation de l'urbanisme, Paris

Lecroart P (2007) From the plan to the city: managing large-scale urban development projects in Europe. CAHIERS-Institut d Amenagement et d Urbanisme De La Region d Ile de France 146:110
Lecroart P, Palisse J (2007) Large-scale urban development projects in Europe: what lessons can be learnt for the Ile-de-France Region. CAHIERSInstitut d Amenagement et d Urbanisme De La Region d Ile de France 146:5
Lecroart P, Palisse J-P, Solupe M (2007) Conduire le changement dans les métropoles. IAURIF, 146
Loraux J (1946) Caen, la ville d'hier et de demain. L'information géographique 10(5):184–189
Lucchini F (2012) La fabrique des lieux: réapproprier des lieux par la culture. Édité par Université de Rouen-Atelier 231-La Friche-UMR CNRS IDEES
Lupi T (2007) IJburg, les hauts et les bas d'un nouveau quartier (sub) urbain d'Amsterdam. Cahiers de l'Institut d'Aménagement et d'Urbanisme de la Région d'Ile-de-France 147:126–134
Miller L (2011) Sustainable waterfront revitalization. Édité par Faculty of California Polytechnic State University San Luis Obispo
Oakley S, Rofe M (2008) Global space or local place? The port Adelaide waterfront redevelopment and entrepreneurial urban governance. The University of Adelaide, Adelaide
Pecqueur B (2005) Les territoires créateurs de nouvelles ressources productives: le cas de l'agglomération grenobloise. Géographie, Economie, Sociétés 7(3):255–268
Pecqueur B (2006) Le tournant territorial de l'économie globale. Espaces et sociétés 124–125(2):17–32
Pecqueur B (2007) L'économie territoriale: une autre analyse de la globalisation. L'Économie politique 1(33):41–52
Pinson G (1985) L'espace social d'une bourgeoisie urbaine: Caen au début du XXe siècle. Villes et sociétés urbaines: Basse-Normandie, XVIe et XXe siècle. Cahier des annales de Normandie 19:205–235
Prelorenzo C (2010) Le retour de la ville portuaire. Les cahiers de la Méditerranée 80:157–167
Raoulx B (1996) Cherbourg et Caen: deux modèles de relation entre la ville et le port. Norois 169:93
Raoulx B (1999) La pratique de l'espace urbain des populations marginales: l'exemple des «binners» de Vancouver. Espace Populations Sociétés 17(2):353–357
Rivoallan Y (2013) Le passé maritime de Caen et influence sur son aménagement aujourd'hui. Thèse professionnelle soutenue à l'EM Normandie
Salet W, Halbert L, Bragg A (2007) Conduire le changement dans les métropoles. IAURIF, 146
Savage VR, Huang S, Chang TC (2004) The Singapore River thematic zone: sustainable tourism in an urban context. Geogr J 170 (3):212–225
Soraya M (2009) Study on urban waterfront Development for sustainable tourism in coastal town: the case of Pujabda Bay in the municipality of Mati. Davao Oriental
Tineke L, Forsyth D, Van der Want J (2007) Conduire le changement dans les métropoles. IAURIF, 146
Yassin A, Razali M, Adnan Y (2013) Examining governance for waterfront development in Malaysia. In: Proceedings of the 10th AAM international conference 2013, Malaysia
Zhang H, Shen X, Gao J (2011) Spatial structure of the leisure zone in urban waterfront: a case study of the Grand Canal in downtown Hangzhou. Urban and Rural Studies 30(10):1891–1900

Bordeaux's Playful and Sporty Maritime Life: A Revolution of Venues and Activities

Jean-Pierre Augustin

Summary

In Bordeaux, the succession of two different types of maritime life, one based on work and trade, the other related to leisure activities and sports, has been leaning increasingly since the 1960s in favour of the second. This movement is part of the transformation and reappropriation of the historicity of river towns and their recomposition. The practices diversify: the quays are landscaped into leisure and sports areas; the banks, more and more attractive, become public spaces with the open sea as the horizon; the river and its surroundings are no longer just décors or workplaces, but offer themselves as facilities for recreational pursuits. The spaces change name and purpose: wet docks become marinas, the river site becomes yachting basin, the hangars become venues for recreational or commercial meetings, the shipyards seek new functions, even the new bridges are largely for pedestrians, cyclists and people enjoying themselves. The incessant invention of new practices and the transformation of places bear witness to the innovation and transformation capacities of an urban and post-industrial society seeking a new sense of existence through leisure, by mixing an ideational world of representations with the real world of places in another economic, political and participatory organisation.

Keywords

Bordeaux · Maritime life · Sports · Leisure activities · Landscaping

Introduction

Bordeaux is a city whose economic growth was largely founded on maritime trading. A market town during the high Middle Ages, it turned towards northern Europe under British influence, then towards the colonial goods coming from the "islands", which made its fortune in the eighteenth century. The first colonial port of the realm before the Revolution, the urbanistic transformations of the town testify to the match between maritime function and urban organisation. Three centuries later, the context has fundamentally changed; the town centre has become a gigantic, open-air museum that seeks to valorise the assets of its maritime past by integrating them into contemporary enterprises and projects. Nautical developments, transatlantic races, large sailing ships, right bank façades, cruise stopovers, nautical trade fairs and specialist colloquia, are so many initiatives promoting a new maritime life in which history, practices and representations of the sea are intertwined.

The nautical heritage is reused by the city dwellers' current ideas of the sea. The succession and intertwining of two different types of maritime life, one based on work and trade, the other related to leisure activities and sports, are part of the reappropriation of the city and its recomposition.

Maritime Life Founded on Harbour Work and Trade

Bordeaux has historically benefited from several periods of expansion, which have been likened to a "golden age": the Gallo-Roman period, the English medieval period, the French period of the eighteenth century and the industrial and colonial period. The last two periods, through their urban

J.-P. Augustin (✉)
Université Bordeaux-Montaigne, UMR Passages/CNRS, Pessac, France
e-mail: Jean-Pierre.Augustin@msha.fr

C. Machemehl et al. (eds.), *Reclaiming and Rewilding River Cities for Outdoor Recreation*, Estuaries of the World,
https://doi.org/10.1007/978-3-030-48709-6_12

planning and monumental architecture aims, have left the deepest traces and marked the area of the town.

In the eighteenth century, the stewards of the realm and knowledgeable architects enabled the city to emerge from its medieval straightjacket and adorn itself with the signs of modernity. Sent on a mission by Louis XV in 1729, Jack Gabriel, the King's first architect, fell under the charm of the commercial bustle of the town and the bend of the river. In 1755, Place Royale was inaugurated with its two buildings on the edge of the quays, the Ferme générale and the Trade Exchange. The style of architecture was French, with high arcades on the ground floor, then two stories and gable roofs covered in slates. Facing the port de la lune where hundreds of boats anchored, the uniform lines of the façade of the quays stretched for more than a kilometre. Bordeaux at the time was the leading colonial port of the country with trade oriented towards northern Europe, Martinique and Santo Domingo, but also towards the coasts of Africa, America and Asia. The traffic was diversified: exports of wines, manufactured goods, deep-sea fishing and "comestibles" assured the growth of Bordeaux trading. This golden age was thrown into difficulty by the Empire and the naval blockade (Crouzet 1968).

Trade recovered gradually and in 1860, Bordeaux was once again the third-ranking French port for maritime traffic after Marseille and Le Havre, followed by Dunkirk and Rouen; it remained in third place until just before the world war. This situation is the result of free trade with Europe and the opening towards other continents. Half the traffic was with northern Europe: industrialisation made Bordeaux the port of arrival for coal from Wales, which was supplied with wooden posts for the mines from the Landes forest. Trade with the Americas remained brisk; while it declined with the Caribbean, it grew with South America to a quarter of all exchanges. Bordeaux trading also extended to new spaces in Africa and the Far East. From 1896 to 1914, the harbour facilities were modernised: the quays were built to facilitate moorings, docks were constructed for storing the merchandise and cranes were installed for loading. Faced with the growth in traffic and ambitious development projects, the construction of an outer harbour at Le Verdon was planned from 1914. The war changed these plans and Bordeaux, with the port of Bassens, became the major landing base for American troops.

After 1920, the infrastructure works, and in particular the extension of the downstream port facilities, continued. Coal, grains, phosphates, oilseeds, wood and sugar represented the major part of the imports. Exports were more limited due to the low level of industrialisation in the hinterland (Lajugie 1972). Fifty-five regular lines connected Bordeaux to the rest of the world and the principle of autonomy of the port was decided in 1924. Colonial trade, in particular with sub-Saharan Africa, had pride of place in the commercial exchanges. Passenger traffic on liners such as the Lutétia, Massilia, Brazza, Foucauld, Meknès and Marrakech grew spectacularly to reach the record figure of almost 60,000 in 1949 (Lebegue 1988). The Bordeaux international and colonial fair set up in the month of June, Place des Quinconces, affirmed the link with overseas by building pavilions for each group of colonies. At the same time, Bordeaux became a home port for deep-sea fishing, while the wine trade was growing steadily.

After the Liberation, the economic recovery of the region was related to the maritime recovery. Traffic resumed slowly, the lines were re-opened, but the pace of growth in trade was less sustained than in the other French ports and in 1954, Bordeaux lost fifth place to Nantes-Saint-Nazaire. The movement of merchandise—with four million tonnes—was close to the 1938 figure, but the composition was considerably different with the progress of oil. Hydrocarbons reached eight million tonnes in 1981, then decreased with the closing of the refineries. In 1990, the autonomous port was the sixth largest in France, with nine million tonnes of freight, four of which were hydrocarbons. It stretched across a hundred kilometres, from Bassens, the estuary port, to Le Verdon, seaport. Bassens possessed high-performance tools for bulk, industrial goods and foodstuffs, miscellaneous goods and wood. Le Verdon is a terminal open day and night, accessible to the larger ships as well as to the third-generation container ships. An investment programme aimed to boost competitiveness. But in its recent evolution, it is the disappearance of most visible maritime activities in the town that should be remembered with, in particular: the suppression of the major cruise liners in 1957, the displacement or removal of the industrial harbour infrastructures; in 1970, the closure of the last major shipyard which had employed up to 6000 workers and fed an important sub-contracting activity; the end of the Terres-Neuvas deep-sea fishing expeditions and the abandoning of Place des Quinconces for the international fair.

The decline of maritime life based on work and trade in Bordeaux, as in the other estuary ports of Europe, is related to a series of economic mutations. The collective memory of the town and the port wastelands retain the souvenir of this maritime life of the past which left its mark on the urban landscape. In parallel, the elements of a new maritime life based on leisure activities were affirming themselves and interfering with the Bordeaux people's practices and representations of the sea.

Maritime Life Based on Sports and Leisure Activities

The strongest symbolic change was the landscaping of the banks of the Garonne to allow the city to finally turn towards the river and be free once more to imagine the great overseas. Almost 4.5 kilometres of banks were opened from the Saint-

Fig. 1 The water mirror, at the edge of the river

Jean Bridge to the wet docks, offering the passers-by the possibility of owning the spirit of the place. Major names in architecture and landscaping, Portzamparc, Villemotte and Corrajoud were mobilised to achieve the connection between past and present and offer this mixture of culture and nature, mineral and plant. The culmination of this cultural landscape is the "water mirror" (Fig. 1), installed opposite the mythical Place de la Bourse, using innovative technology to reflect and mingle the eighteenth century façades and the Garonne. Inaugurated in 2007, it has become the emblematic attraction of the town, where children play with the water while amateurs artists offer up their creations.

But the river is no longer simply a décor open to the high sea, it has become a facility for activities, beyond the "whaleboats" of the sea scouts, the start of nautical races (Fig. 2) the reintroduction of regattas, the landscaping of marinas, mini-cruises and the major project of the wet docks are all witness to this (Augustin 2006). One of the strong demonstrations of this reappropriation is the return of the swimming event across the Garonne, which had been a great success up until the 1950s. The idea was proposed by the Girondins omnisports club in 2007 with the participation of Olympic champion Jean Boiteux and the president of the club Patrick Baqué. Since this date, each year close on 500 swimmers set off from the Parlier pontoon for a major fete mixing competitive swimmers and amateurs swimming for fun (Fig. 3). Marc Lafosse, the organiser of the cross river swim and leader of the project for tidal turbines under the stone Bridge, has plenty of ideas for the people of Bordeaux to really reinvest in the river, the association "*du port de la lune*" already provides for demonstrations around swimming, canoeing-kayaking, rowing and sailing... (Bessy 2014).

The activities diversify (Lefebvre et al. 2013), the quays are landscaped into leisure and sports areas; the banks, more and more attractive, become public spaces (Augustin 2001) with the open sea as the horizon; the river and its surroundings are no longer just décors or workplaces, but offer themselves as facilities for recreational pursuits. The spaces change name and purpose: wet docks become marinas, the river site becomes a yachting basin, the hangars become venues for recreational or commercial meetings, the shipyards seek new functions, even the new bridges are largely for pedestrians, cyclists and people enjoying themselves (Augustin 2002).

Sports enthusiasts do not only practice on the playing fields, stadiums and gymnasiums. In Bordeaux as in other maritime cities, they are increasingly investing public spaces and in particular the parks, squares and waterfronts where in

Fig. 2 Start of a Trans-Atlantic race

various different forms, runners, rollers, skaters and bikers appropriate the areas intended for other purposes. This phenomenon helps to create a flexible urbaneness that allows people the possibility of accessing diversified places, entering into relation with a variety of groups and of taking part on occasion in scheduled or spontaneous meetings. These new forms of use pose the question of their organisation and management as well as of their relation to the facilities. Places to see, to be seen, to mix with other people, public spaces allow an interactive play where the roles of the players and spectators are interchangeable. The experience and perception of these places remains, however, complex and the appreciation of the citizens with respect to them can vary depending on the individual.

A phenomenon that is particularly visible in cities with a seafront, estuary or river, the new practices correspond to the fast distribution of activities organised more by lifestyles than by sports institutions. They are a part of an urban vision where the concepts of mobility, proximity and connection are joined by the evolution of the private and public, of the individual and the community.

The changes have come from internal transformations, particularly from the valorisation of the body by the media, and technological inventions bringing onto the market a supply of products for individual practices. The *skateboard*, a small plank of wood with wheels that allows you to move around and make figures on the urban furniture, appeared in 1962; then *in-line-skates*, resulting from a transformation of roller skates, made an astonishing breakthrough in the following decade. The use of cycles, in the form of *bikes,* grew rapidly, with the special feature of making use of a means of transport for various activities: the bicross with small wheels and large handlebars to facilitate figures due to the lowering of its centre of gravity, the BMX that absorbs more violent shocks and the mountain bike for the twin uses of acrobatics and ambulation. These machines became very popular with young people and form the core street sports to which should be added streetball and street hockey, which, although requiring transient teams to be built, are a part of the same rationale. Carried by the multiple offers of the major sports equipment firms, these practices find, in harbour towns, in particular, outlets and favour the emergence of urban territorialities for athletes, which fit into the gaps between town and riverbank.

The urban beaches along the Bordeaux watercourses and interior bodies of water were organised from 1990s onwards. During the summer, Bordeaux-Lake or Bègles attract the

Fig. 3 Swimming across the Garonne River

citizens of Bordeaux who resist the attraction of the coast or who cannot afford to go there; the sailing schools encourage the apprenticeship and practice of sailing and Bordeaux-Lake remains emblematic of nature and water sports in the town. But the novelty comes from the water sports complexes and the diversification of practices in a single place. The range of facilities on offer grows, traditional swimming pools are recycled and rehabilitated with the addition of waterslides, wave pools, spaces with circulating currents and new, ever more sophisticated concepts are proposed, adjoining sports activities to retail units and meeting places. The watersports offer becomes multi-form. Thus, the *Calicéo* centre built at the beginning of the 2010s in the Tasta neighbourhood of Bruges is a well-being and aquatic leisure centre of 2700 square metres; proposed by architect Bertrand Hubert, it has an outdoor pool open all year round with water at a temperature of 33 °C, completed by geysers, saunas and steam baths that can accommodate 400 locals and tourists at the same time.

On the right bank, the project of the Lormont relaxation and leisure water centre, *Cascades de Garonne*, designed by architect Jean-Michel Ruols, proposes a water park of 9 hectares with 30-metre waterfall and lagoon; it is expected to be accompanied by economic activities and a housing programme, and could welcome up to 50,000 visitors a year. Presented as an emblem of a reviving right bank, it is still seeking funding.

The tidal bore that follows the first wave of the rising tide every day, creating a wave that can be up to a metre high depending on tidal factors and the morphology of the river, becomes the occasion of new games and sports; a few kilometres from Bordeaux, at Saint-Pardon on the Dordogne, this is an event for hundreds of surfers and thousands of spectators: the *Mascarock* (sixth edition in 2014) has become a ritual celebration with brass bands and concerts. Lastly, less than an hour from Bordeaux, Arcachon Bay with its nautical fetes and marinas, the surf resorts around the promising waves, of which Lacanau is the symbol, are extensions of the city (Augustin 2000, 2007a, b).

Conclusion

The souvenir of a maritime life based on work and trade remains strong in Bordeaux, all the more so as the abandoned maritime areas mark the town with their footprint. New representations have succeeded the old and an outdoor maritime life related to leisure activities and watersports have been added to the first. They mingle today, and the recent styles of existence and sensitivity generate new meaning. The development of the banks of the Garonne, the organisation of transatlantic races, the construction of marinas correspond, from a common maritime heritage, to the staging of new representations, places and practices. After being deprived of access to the river for almost a century, the people of Bordeaux are reclaiming both banks, which are at the centre of the urban plans of Bordeaux in the twenty-first century. At last the historical spaces that the industrial revolution had differentiated and opposed have been reconnected. The town with its specific features, its past as a major Atlantic port is a perfect example of these mutations that are both concrete and symbolic.

Bordeaux, a sporting town, has not yet quenched its thirst for water and has not taken full advantage of its relation to the river, but there is no lack of projects around participatory environmentally friendly sports events, and everything suggests that the recreational sports activities on the Garonne have only just begun.

References

Augustin J-P (2000) La diversification des loisirs sportifs sur le littoral médocain. Les cahiers espaces 66:68–78

Augustin J-P (2001) Espaces publics et cultures sportives. Géocarrefour, Revue de géographie de Lyon 1:27–30

Augustin J-P (2002) Corps sportifs en ville. Revue Urbanisme 325:38–40

Augustin J-P (2006) Bordeaux, au rythme des cultures sportives. Sud-Ouest Européen 22:127–140

Augustin J-P (2007a) L'attractivité plurielle d'une station océane: Lacanau-océan dans le sud-ouest de la France. Téoros, revue de recherche en tourisme (Montréal) 26(2):39–45

Augustin J-P (2007b) Sports et géographie, spatialités contemporaines et mondialisation. Armand Colin, Collection U, Paris

Bessy O (2014) L'innovation dans l'événement sportif. PUS, Voiron

Crouzet F (1968) La croissance économique de Bordeaux. In: Bordeaux au XVIIIe siècle. Fédération historique du Sud-Ouest, Bordeaux, pp 191–286

Lajugie J (1972) L'essoufflement de la vie économique. In: *Bordeaux au XXe siècle*. Fédération historique du Sud-Ouest, Bordeaux, pp 115–194

Lebegue A (1988) Bordeaux des paquebots. Sud-Ouest, Bordeaux

Lefebvre S et al (2013) Les nouvelles territorialités du sport dans la ville. PUQ, Québec